UFO In

The Age

Declassification & Disclosure

What the Government Knows about UFOs

Based on 30 years of Research and Investigation

Richard R. Lang

Edited by Karen Wuenstel

www.LangPublication.com

Any dialog in this book, reference of any kind, or conclusion made by the author, should **NOT** be associated with MUFON, BAASS, OPUS, ASSWAP, or any other UFO research organization or group, as any official statement or a public position of that organization.

Date: Revised Feb. 2024
ISBN: 9780988360693
LANG PUBLICATION
Printed in USA

Table of Contents

About the Author

Richard Lang

Ufologist - Researcher - Investigator - Author - Speaker

In 2009 Richard managed the **BAASS/MUFON SIP Project** that was facilitated at Bigelow Aerospace and was operated and funded by the Defense Intelligence Agency through the **Advanced Aerospace Weapon Systems Applications Program (AAWSAP)** at the Pentagon. At the time it was the most advanced funded rapid response UFO/UAP Investigative Team in the world.

He is an FAA Licensed Commercial Pilot with a Multi-Engine Rating and has a Bachelor Science Degree in Aeronautics from Embry Riddle Aeronautical University in Daytona Beach, Florida.

Mr. Lang enjoyed a successful career with more than 20 years in the corporate world as a Senior Vice President working in Brokerage and the Trust Investment Division of Commercial Banks.

In the post 911 period he left his banking career to serve as a Special Deputy for the U.S. Marshals Service working as a LEO (Law Enforcement Officer) in the airport.

He worked with the Transportation Security Administration (TSA) as a liaison responsible for implementing federal procedures and dealing with regulatory initiatives associated with aviation security. This also included responsibility for airlines and airport authorities as well as all law enforcement agencies in Virginia that respond to airports during aviation emergencies.

He was a member of the Anti-Terrorist Advisory Council Board (organized by the U.S. Attorney's Office) and frequently served as key speaker on terrorism at quarterly federal regional meeting.

He belonged to the UFO organization known as MUFON and was certified as a Field Investigator; he eventually was appointed as Chief Investigator for both Virginia and North Carolina. He was one of the original six members of the **MUFON STAR** team where he functioned as coordinator and manager of the team. He is currently a MUFON Benefactor and Life Member.

Richard is on the **Advisory Board of OPUS** which is a global organization supporting people having anomalous experiences such as extraordinary states of consciousness, spiritual, parapsychological phenomenon, close encounters with non-human entities and or UFO activity.

Aerospace Research Operations

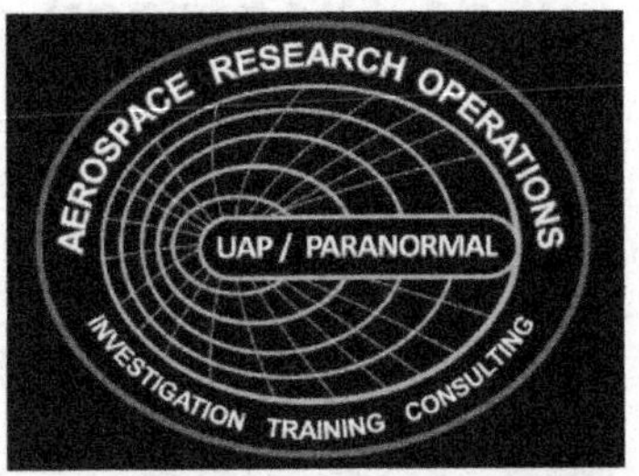

Richard serves as the Principal Partner, and Director of Operations at Aerospace Research Operations This is a recently established scientific research cooperative that is focused on the Unidentified Anomalous Phenomena and the associated high strangeness and paranormal effects that typically result after the encounters.

In 2008 he was showcased as the Lead Investigator in the Discovery Channel TV series **"UFOs over Earth"** and later in 2012 with Canada's Discovery Channel TV series **"Close Encounters."**

He has been interviewed numerous times on podcasts and radio shows including three times on ***Coast-to-Coast AM*** with George Knapp.

He currently serves as owner and editor at:

LANG PUBLICATION

www.LangPublication.com

Lynda Thompson

Research - Technical Advice

Special thanks to Lynda Thompson who has helped me with research and technical data included in this book. Lynda has an educational background in accounting and law. She was a certified State Bar Paralegal in Texas. She worked in several law firms in the areas of litigation and general law practice.

Lynda's skills as a paralegal over many years provide the basis for her investigative research and writing. She has a personal background in UFO/UAP research. She has been a member of MUFON for many years. She is currently the Assistant State Director for Virginia MUFON and has served as the chief investigator in the training of investigators. She has completed numerous UFO investigations as an F3-Advanced Certified Investigator.

Over the last 10 years Lynda has been extensively involved in abduction research working directly with experiencers and organizations that support abductees.

She is an accomplished photo analyst and handles private research for agencies and corporations. Lynda has assisted international speakers, writers, news reporters and scientists in their work. She currently serves as the Director of Research for **Aerospace Research Operations**.

Preface

This book is for the general public, who may be interested in the UFO subject but may not know a lot about it. Unlike my first book, which was written to provide advance technical training for UFO investigators and individuals who have been involved in UFO research for many years, this book is written in general terms to permute public understanding.

In 2023 as this book was being revised for publication, Congress has been holding active nationally televised hearings interviewing Pentagon officials about UFO events. Whistleblowers are testifying in Washington, D.C. before Congressional assemblies about non-human bodies and spacecraft of extraterrestrial origin that have been recovered by government agencies and defense contractors during the last 70 years. The mainstream media news channels have picked this up and talked about it extensively, which has caused a dramatic increase in public awareness about the subject. (more about this in Chapter 8).

Over the last 4 years the Pentagon has started to release information about the government's involvement with UFO activity. They have released numerous photographs and videos taken by navy fighter jet pilots showing close encounters with UFOs that are clearly not from this planet.

75 Years of UFO History

If you are a newcomer interested in UFOs, you need the history lesson to understand the whole picture. The UFO story is not at all new, although many recent newspaper reports and public disclosures, which have been released over the last four years would lead most people to think that this is something new that just recently started to emerge. You need to understand that in fact this is not the case!

The first part of this manuscript was created to walk the reader through a series of historical events relating to UFOs, that started in back in the 1940s, and lead to development of highly classified secret research projects over the last 75 years. Hopefully this will provide some insight into how the secrecy surrounding UFOs started and evolved over the years.

Next looking at the covert government operations and how they classified and manipulated UFO technology, working closely with defense contractors, ultimately keeping it hidden from congress as well as the American people. I think it is important for the public to understand how much has happened over the last 75 years involving secret research and advancements in technology that used funding from covert government tax sources and allowed the Military Industrial Contractors, through the use of misplaced power as President Eisenhower said, to have total control and make billions of dollars for their own interests.

Tides are Turning

I believe we are coming to a critical point where the greatest mystery in human history is about to be unveiled through a slow and systematic release of information and disclosure about UFOs.

Over the last four years there's clearly been a shift in the direction the Pentagon has taken in releasing information associated with UFOs and disseminating it to the public.

In this book I have tried to summarize most of these current information releases and leaks including the newspaper articles, public disclosures and videos of UFOs that were released by the Pentagon.

Next analyzing the recent UFO Assessment Report which was mandated by Congress and just released to the public in 2022. Also looking at the National Defense Authorization Act (NDAA) of 2023 which protects whistleblowers who testify to Congress.

Finally in the last chapter we are speculating on the steps that will be taken in the near future to disclose the full truth about the UFO enigma.

30 Years of UFO Research

Over the last 30 years I have been heavily involved in this research and it's always been interesting to me that so many of the people that I meet and know from my social interactions simply do not understand this subject. Some merely do not take it seriously, some talk about conspiracy theories and fairy tales, others are basically not comfortable because they believe it conflicts with their religious convictions and for a few people it just scares the hell out of them.

Green Monkeys

I was asked to speak at a Rotary Meeting in Charlotte, NC after one of my Discovery Channel shows aired in 2007. At that time, I was a Rotary Member where several hundred of us would typically attend a weekly luncheon meeting. I had put together a presentation that was fairly generic in that it did not include a lot of details about high strangeness, but more information concerning general sightings and percentages of people that were filing these reports.

As I was concluding my presentation one of the participants in the audience stood up trying to make fun and began talking about monkeys that had been spray painted with green paint and turned loose, some people in the meeting started to laugh which embarrassed me a little bit, though I have developed a tolerance for those kinds of disruptions during my presentations.

Airforce Colonel at the Rotary meeting

As fate would have it, after I concluded my presentation one gentlemen in the audience came up to me and told me that he wanted me to meet with his son, who was an Airline Captain. He felt they had something important to tell me about. I knew this gentleman was a retired Air Force Colonel and had been a fighter jet pilot.

A couple days later I met with the colonel and his son, who told me that he was flying as a captain on an airline flight when he had a very close encounter with a UFO. He told me an amazing story about a huge unconventional object that was directly in his flight path and how he made turns to avoid it, while it continued to keep pace and follow the flight for 20 minutes.

Military Intelligence Officer

A month later I had a similar experience with another officer of the bank where I worked. We were having lunch together and he had apparently watched one of my Discovery Channel shows. He said something spontaneous to me like you know this UFO stuff is real!

I acknowledged it and asked what do you mean? I had known this guy for several years as he was the manager of one of the biggest branches in our bank network. He was a very serious straight up no nonsense kind of guy who would never lie about anything. He confided to me that he was an army officer training in an intelligence program (1970s) on a base somewhere in North Carolina (Fort Bragg). He said that he had never shared this story with anyone before because he was afraid of being criticized and ridiculed.

He was apparently in a barracks sleeping at about 1:30 AM in the morning when a general alarm went off and officers came in, rousted everybody up in his unit and transported them to a beach which at the time did not make sense to him but he was not about to question the officers in charge.

It was somewhere around 2:45 AM when they arrived at the beach. He said that he along with the other members of his unit were standing milling around on the beach, not sure why they were there. Then slowly overhead they noticed that there was something huge in the sky that was slowly descending over the beach where they were standing. He estimated that it stopped 1000 feet altitude above them and did not make any sound. He stated it was so huge it blocked out the entire night sky and stars. I asked him to define huge and he said about the size of a small strip mall shopping center. He described the object as being black with a lot of very intricate details on it and a lot of multi-colored lights twinkling all over it, kind of like looking down on a big city from a great distance high in the air.

After a few minutes the object started to rise up in the sky. He said it silently went straight up and kept going until it vanished out of sight. The guys in his unit were taken back to their barracks when several high-ranking officers came in and told them "This never happened" and that they were never to talk about it again.

Social Interactions as a UFO Researcher

So, my social experiences over the years have been a lot like that, typically when someone learns about my UFO research, either they start up about the green monkey nonsense to make fun, or they get very serious and take me aside to tell me about an experience they had involving some kind of close encounter with a UFO.

Public Opinion Polls

It is interesting to note how the public in general feels about UFOs and extraterrestrials. In 2009 AOL conducted a poll that involved about 500,000 Americans. The participants in the poll were asked two questions:

1) Do you think that Extraterrestrials have visited Earth?
    - 73% answered YES.
    - 17% answered NO.
    - 10% were undecided.
2) Do you think the government is trying to cover up the existence of Extraterrestrial life?
    - 76% answered YES.
    - 15% answered NO.
    - 9% were undecided.

Over the last year or so these numbers have dramatically increased to about 90% of the people now answering yes to the two questions.

This is due in part to the fact that the $2.3 trillion dollar Appropriations Bill that was signed into law last December mandated a report detailing what the government knows about UFOs being released and declassified that year.

With that in the works, many of the mainstream media news outlets and politicians are now talking about UFOs. Newspapers and TV channels have been showing videos of UFOs that were taken from fighter jet cockpit cameras and interviewing navy pilots, who are now coming forward talking about the UFO subject.

We're going to talk a lot in the next few chapters about that information, by looking at the steps that might be taken to disclose and declassify UFOs as well as the effects that action may have on our society and economy.

Chapter 1 - Introduction

This is an introduction to the subject of Ufology, starting with definitions of the most common terms referring to or describing UFOs. Then a very brief overview of UFO research and some historical references about Extraterrestrials including people who have had close encounters with entities that appear to be non-human.

Chapter 2 - 30 Plus Years of Research

I wanted to include a little bit about myself and the 30 plus years of research and investigation that I have conducted. Beginning in 1977 with an early experience I had while flying alone at night in Florida involving in a UFO encounter with airline jet aircraft.

At the risk of being slightly narcissistic, I included a little bit about my education, background, investigative experience, and my credentials to make sure that the readers understand that I know what I am talking about.

Chapter 3 - Government Cover-up and How it all Started!

Anything to do with UFOs, Aliens and Extraterrestrials is the most highly classified and carefully guarded secret in the U.S. Government. In this chapter we will take a look at some history to understand how this all started and what evolved, including some interesting developments during the Apollo 11 moon landing. We also explore the concept of secrecy that has been applied to protect US government technological research and back engineering projects that have worked on recovered UFO crash wreckage.

Chapter 4 - Covert Government Operations is about the secrecy surrounding the UFO/UAP phenomena and covert back engineering programs, we also look at how the Media is controlled in keeping the secrets.

Chapter 5 - DIA - BAASS - MUFON - ASSWAP Behind the scenes in 2008 the Defense Intelligence Agency (DIA) was secretly preparing to fund several UFO/UAP research programs which were ultimately associated with MUFON through the Pentagon's **Advanced Aerospace Weapon Systems Applications Program (AAWSAP)**. I worked on part of this project which will be explained in this chapter.

Chapter 6 - Problems Associated with Disclosure and Declassification looking at the adverse consequences of disclosure including how it will affect the Stock Market. Also looking at the very high strangeness associated with UFO encounters and the difficulty of getting the public to understand it.

Chapter 7 News Events and Information Releases. Looking at the UAP taskforce and other developments from the pentagon and organizations involved in leaking and releasing information to the press.

Chapter 8 The Assessment Report of 2021 and The National Defense Authorization Act (NDAA) of 2023 This 9-page Assessment Report really didn't provide a lot of useful information about UFOs, however, it is the first step in a long process that may ultimately lead to Full Disclosure. The NDAA provided that people with information about UFO / UAPs have been given amnesty and immunity from prosecution while testifying to Congress.

Chapter 9 - Steps to Full Disclosure Could it be that this initial disclosure report is part of a longer-term disclosure project to change the public awareness and ultimately disclose the full truth about UAPs and extraterrestrial origin?

Richard Lang

Palmyra, Virginia USA

(Written 8 miles from Monticello)

Chapter 1
Introduction

Ufology is the study of **UFOs** (Unidentified Flying Objects) and especially the investigation of recorded sightings and human encounters in the current sphere of independent research.

The term **UAP** (Unidentified Anomalous Phenomena) previously known as (Unidentified Aerial Phenomena) has evolved because we now understand that these objects simply do not fly in the conventional sense, although they occupy airspace above the Earth, as such the phenomena has little to do with aerodynamics or conventional flight.

Naval aviators and military air traffic controllers use the term **AAV (**Advanced Atmospheric Vehicles) also known as (Anomalous Aerial Vehicles) as opposed to UFO or UAP.

The term **"Fast Walker"** is often used by fighter jet pilots on the radio to identify a UAP that is appearing at high velocity in close proximity to their jet aircraft.

A **USO** (Unidentified Submerged Object) is an unidentified object that is submerged underwater (more recently known as "Unidentified Aerospace-Undersea Phenomena). The U.S. Navy and U.S. Coast Guard have had their share of encounters with these types of objects. These USOs have been detected and tracked on sonar traveling at unbelievable speeds underwater (more than 300 MPG).

Shapes of UFOs

The boomerang shaped UFO pictured here has been estimated to be over two miles across in size.

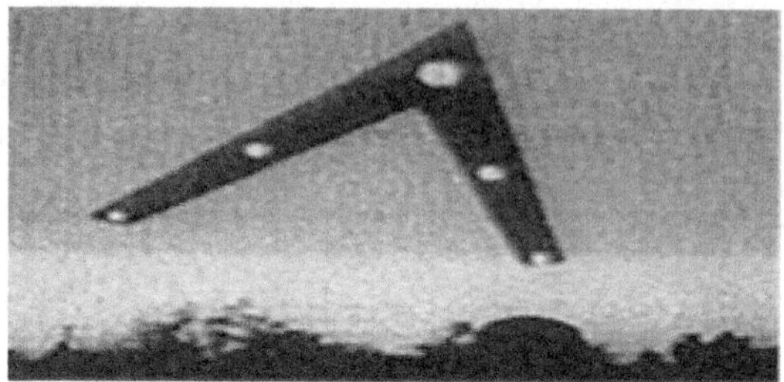

Witnesses have described the shapes of UFOs varying from a sphere to a glowing tube. Some resemble flying saucers.

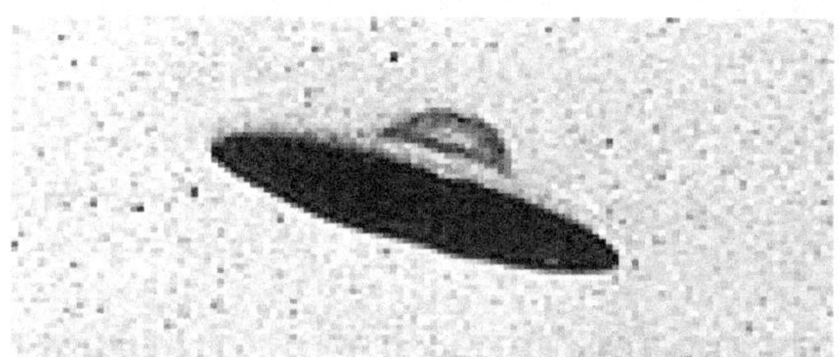

Others are triangle shapes as illustrated here. Some triangle vehicles have been estimated to be the length of three football fields and three stories high.

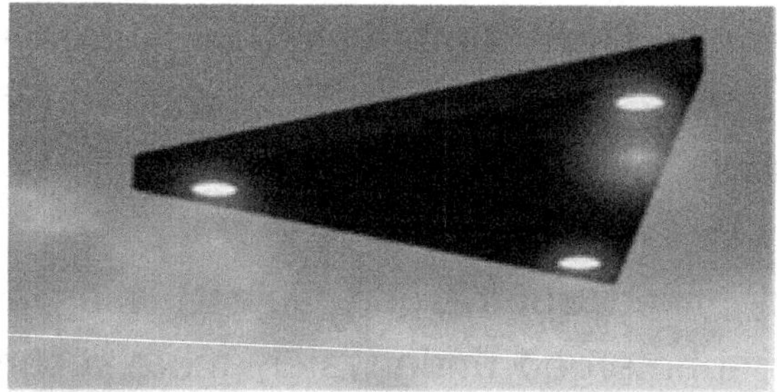

Very common are metallic cylinder shapes that may be more than 300 feet long. There are many reports of ball shaped orbs with reddish-orange glow or fire-like discharges ranging in size from a soccer ball when close to the ground to much larger orbs at higher altitudes. Sometime orbs are reported in blue-white and bright white colors of various sizes.

Unusual Trajectory Flight Characteristics

Incredible speed and maneuverability that is not attainable by conventional aircraft are commonly observed in most all cases, to include hovering, instant acceleration (0 to 1000 miles per hour), stopping, sudden reversal of direction, high speed abrupt directional changes and 90 degree turns.

Appearance: The outer skin is often metallic-looking and seamless. At night, a glow or protective field layer can be observed around some crafts. Sometimes color changes are observed with changes in speed.

No visible means of propulsion: There are no observed propellers, jet engines, rocket thrusters, rotors, wings, rudders, elevators, or flaps that would be associated with conventional aircraft. There is no visible exhaust, vaper trail, or heat signatures trails that show on passive infrared cockpit cameras.

Multidimensional: Crafts that are reported to disappear and reappear may be shifting in and out of another dimension.

Anti-gravity (electrogravitic) propulsion

Extraterrestrial spacecrafts have a system of propulsion that's commonly known as electrogravitic. The mechanics of these propulsion systems are housed in a sealed sphere made of soft silver-white metal. The metallic sphere is filled with mercury. Inside the sphere is a copper circular device that rotates within the mercury creating an electromagnetic field and an anti-gravity bubble inside and around the craft.

The antigravity field repels the craft from the Earth similar to the effect of placing poles of two magnets together. The rotational speed of the rotating copper device controls the power (intensity) of the field and the speed of the craft. This is very similar to the magnetic field produced by the Earth with molten metals at the core which are spinning, uncannily similar to the way motors and generators work.

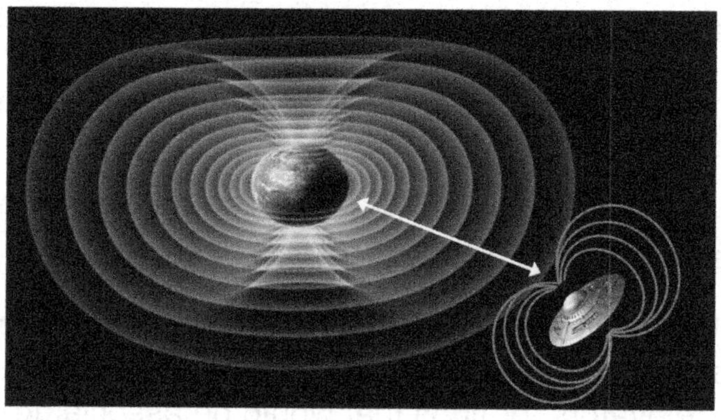

The occupants of the craft have no sensation of speed, inertia or gravity which explains how they can travel at tremendous speeds and make 90 degree turns.

Encounters with Entities

Hundreds of people have observed strange events and had close encounters with entities that appear to be non-human. They fail to report their experiences mostly out of fear, embarrassment or potential to be ridiculed; even more disconcerting are witness reports associated with close encounters that result in physiological effects like marks and burns on the body, various illnesses that result after an encounter and strange psychic effects that occur during and after the encounters. Witnesses typically leave the scenes of these events in shock, disoriented and sometimes sick and in utter disbelief. They often find themselves alone and isolated, afraid to tell their closest friends or even family members what they experienced.

Law Enforcement Not Trained for This

To compound these difficulties, law enforcement officers and emergency responders are regrettably not trained to deal with the high strangeness associated with these events or the public controversy that is so often connected to these cases.

Dr Keith Taylor conducted a Survey of over 400 law enforcement agencies across the country trying to ascertain information about current training and policies for responding to UAP events. Out of the 400 larger agencies contacted, none of them have any established policies, procedures, or training for dealing with UAP encounters.

UFO Reports Around the World

Every day ordinary people from all walks of life and in every part of the world see unusual phenomena in the airspace above this planet. In the United States alone 650 to 850 people report such events on UFO Reporting websites every month. Compound this number several times over if you include South America, Europe, Asia and Middle East. [1]

UFO Reports and Data Bases

With the development of internet databases and automated reporting sites on the internet, computerized reporting systems have developed. Individuals who have a sighting or experience can go online and submit a report.

These report websites have evolved to be quite complex and allow the individual to report in detail the experience by answering questions and filling in blanks on the digital report form. Photographic images, audio recordings and video can be uploaded as well.

The on-line reports feed into a database that can be sorted for a variety of reporting applications. As the database grows over the years the data becomes more useful and valuable. Current research indicates:

- Upwards of over 14 % of Americans have seen what they believe to be a UFO.
- Worldwide sightings of UFOs are estimated to be at 100 million.
- Only 1 in 450 UFO witnesses ever report a sighting.

Reports are Explainable in 80% of the Cases

Many of these events that are observed take place at great distances and high altitudes, often appearing as strange objects or lights in the night sky. Investigators estimate that about 80% of these sightings ultimately will have some rational explanation, such as the unusual position of a planet, a satellite and even the International Space Station as it could be vividly seen from the ground at night. Other explanations are attributed to natural causes such as meteors, comets, reflections, marsh gas and ball lightning.

On the other hand, 20% of these sightings cannot be easily explained, particularly when objects and lights that appear to be at close proximity to witnesses and in some rare cases where objects appear to be on the ground or have some physical effect on the ground.

Classification of UFO Reports

- **CE1 - Close Encounters of the First Kind**
  Visual sightings of a UFO seemingly less than 1000 feet away show considerable detail.
- **CE2 - Close Encounters of the Second Kind**
  A UFO event in which a physical effect is alleged; this can be interference in the functioning of a vehicle or electronic device, animals reacting, a physiological effect such as paralysis or heat and discomfort in the witness or some physical trace such as impressions in the ground, scorched or otherwise affected vegetation or a chemical trace.
- **CE3 - Close Encounters of the Third Kind**
  UFO encounters in which an entity is present, these include humanoids, robots and humans who seem to be occupants of a UFO.
- **CE4 - Close Encounters of the Fourth Kind** is a UFO event in which a human is abducted by a UFO or its occupants.
- **CE5 - Close Encounters of the Fifth Kind**
  A Close Encounter of the Fifth Kind is a UFO event that involves direct communication between aliens and humans.

Authenticity in Reporting UFO Encounters

Reports of UFO encounters are often associated with very credible witnesses including airline pilots, military pilots, astronauts, police officers, fire fighters, military officers, members of Congress and U.S. Presidents including Jimmy Carter and Ronald Reagan. Even Thomas Jefferson wrote about red orbs that were observed in the night sky around Monticello.

Astronauts

There have been many sightings by U.S. and Soviet astronauts. In 1964 three Russian astronauts aboard the Voskhod reported that they were surrounded by a "formation of fast-moving disc-shaped objects”. In 1966, Gemini XII astronauts Jim Lovell and Edwin Aldrin said they saw four UFOs linked together.

During the Apollo II mission astronauts Neil Armstrong, Edwin Aldrin and Michael Collins reported sightings of what were saucer shaped UFOs during the first flight to the lunar surface. (More about this in Chapter 6).

Historical References

There are countless historical references that depict ghostly bodies and supernatural occurrences that share an uncanny similarity to modern day accounts of UFOs and close encounters reports.

Ancient texts are filled with descriptions of flying machines, cave drawings and paintings in pyramids of what appear to be entities in spacesuits and saucer shaped space crafts can be found around the world.

The Baptism of Christ 1710

15th Century Painting

For hundreds of years mysterious objects in the sky and strange moving lights have been reported by many people including the military pilots in World War II who called them "foo fighters".

Extraterrestrial Visitors

The consensus of researchers like me, who have been looking at this data for a long time, is that there are likely about a half-dozen different species of extraterrestrial visitors (ETs) who visit here, and they have different agendas. This would account for the different shapes and types of crafts that are reported in this research and the various body types that have been observed.

Humanoid Body Types

A humanoid body is defined as having two arms, two legs, a head with two eyes, and can walk upright like we do.

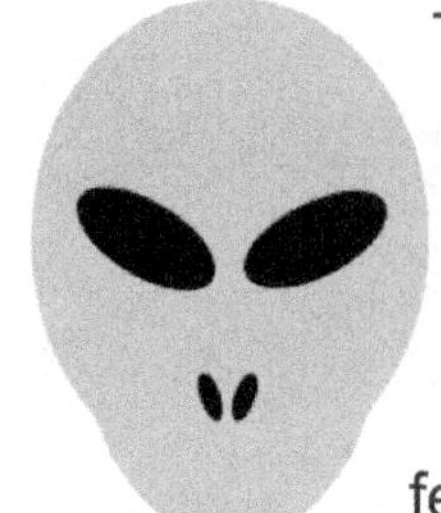

The most common being is a small humanoid that appears very thin, frail, has a big head, large dark eyes and stands about 4 feet tall, they are commonly referred to as Grays; some grays are also reported to stand 7 or 8 feet tall.

Also quite common are beings that look just like us; they appear to be Scandinavian in appearance (blond hair, blue eyes) and are described as very good looking and physically fit. They are seen as males and females and are so normal looking they could blend in in any social environment.

Finally, there are humanoid beings that are more reptilian, or insect-like in appearance, sometimes described looking like a praying mantis which is obviously very frightening to witnesses.

NHI (Non-Human Intelligence) Use of the term NHI has recently evolved particularly with government whistleblowers in reference to biologicals that have been recovered during crash retrieval operations. They are not ready to use the term "extraterrestrial" however they clearly want to indicate that the vehicle wreckage and recovered crafts were not manufactured by humans and the non-human biological occupants of the crafts are not of this planet.

Origin

The questions that are often asked: Do they come from other planets? Do they come from other dimensions? Do they come from other time constraints? The answer may be all of the above.

1. **Other Planets**

There are many other planets and solar systems that exist around the galaxy. There is a lot we don't know about these extraterrestrial entities, but one thing for sure we do know is that their technology is hundreds of years more advanced than ours.

So inner stellar travel might be possible for them in ways that we really do not understand. Our physics defines the shortest distance between two points is a straight line. So, take a sheet of paper and draw a point in two of the opposing corners and then draw a straight line to connect points. Now take the piece of paper and fold it together so that the two points touch each other. So, what if they have a technology where they can bend the fabric of space the way you bent that piece of paper and travel from one point to another almost instantaneously?

2. **Other Dimensions**

Quantum physics now claims that there are dimensions that exist outside this three-dimensional time space reality we live in. CERN (on the Swiss French border) operates the most powerful particle accelerator on Earth, which they use to explore secrets of the universe and see into the other dimensions. They have identified 12 dimensions outside our three-dimensional time-space reality we live in. Clearly, some of these crafts and the ET's who are operating them are somehow able to move in and out of our dimension. [3]

The study of this phenomena is called T**rans-Dimensional Physics**. Very often in case reports, objects are reported to materialize, disappear, and reappear somewhere else. (More about this in Chapter 7).

3. **Time Dilation**

There are numerous reports regarding close encounters with extraterrestrial crafts where individuals report missing time. Clearly with the technology they have, ETs are able to manipulate time. Researchers believe that some of the ET visitors that come here may actually be time traveling from many years in the future.

Abduction Research

Defined as Alien Abduction (sometimes also called Abduction Phenomenon, Alien Abduction Syndrome, or UFO Abduction) refers to the phenomenon of people reporting what they believe to be the experience of being abducted by extraterrestrial beings (ETs) and taken into a ship or spacecraft. They typically allege to have been subjected to physical examination and in some cases psychological experimentation, then they are returned to their original environment. Individuals involved in abductions are sometimes referred to as A**bductees** or **Experiencers**.

MILABS - Military Abductions

In some cases, abductees report the presence of uniformed human military officers as well as extraterrestrial beings in the same spacecraft where the examinations are conducted during the abduction. Although very surprising, I've seen this in a number of reports with witness testimony to that effect.

Classified - Security Clearances

Since 1947 and the first Roswell UFO crash event a policy of absolute secrecy has been applied to protect U.S. technological research and back engineering projects that have recovered UFO crash wreckage. This research is the most highly classified operation and the most carefully guarded secret in the U.S. Government.

Disclosure and Declassification

Anything to do with UFOs, aliens and extraterrestrials including sightings, landings crashes, crash recovery, crash retrieval teams, and research facilities is classified significantly above Top Secret, only those who have the highest security clearances have access to this information.

The concept of "Disclosure" in the UFO community is that the U.S. Government has classified and withheld all information on alien contact and extraterrestrials. Full disclosure and declassification are being pursued by activist lobbying groups, UFO organizations and most recently by members of Congress. Pressure on government agencies to declassify information on UFOs is coming into the public purview. More about this in Chapter 11.

Many believe that there is a time coming soon when the governments around the world will disclose in some form what is known about UFO/UAP Phenomena.

END CHAPTER

Chapter 2
30 Plus Years of UFO Research

For well over 30 years, I have pursued an interest in UFO research, which started out from a very practical mindset, that was the consequence of flight training and a very scientific education in Aeronautics. I remember thinking that this enigma (UFO mystery) was going to be resolved with high-speed camera equipment, powerful lenses, and scientific instruments. For many years, I followed a discipline searching for physical evidence and looking for practical nut and bolt types of answers.

Ultimately I ended up on a journey that opened many doors and introduced me to numerous interesting and gifted people who shared with me their bizarre experiences and told me about encounters involving very high strangeness. All this research eventually resulted with many more questions than answers and left me 30 years later with a much different understanding of physics, physical reality, and human consciousness.

Early Experience as a Pilot

The journey began for me sometime in the early part of 1976 in Florida. I was flying alone in the cockpit of a high performance four passenger single engine aircraft. It was night, well after dark at about 2500 feet AGL (altitude) and I was cruising at about 140 knots heading generally north toward Jacksonville. I was a flight student at that time, working on my Flight Instructor rating and an accomplished commercial pilot. I was attending Embry Riddle Aeronautical

University and was to graduate with my Bachelor's Degree in Aeronautical Studies the following year. [2]

Flying up the East Florida coast at night is a stunning ride particularly if you are offshore and out over the water where you have a view of the beaches and the spectacular lights on shore.

The flight that night started out much like any other adventure with a flight plan originating in Daytona Beach, a scheduled landing in Northern Florida (Jacksonville) and a round robin trip back home to Daytona. Although I had an instrument rating at that time, but I was flying **VFR** (visual flight rules) that evening.

We did not have GPS back then in General Aviation so navigation was essentially facilitated with flight charts and radio navigation equipment known as VOR which was radio equipment designed to track radials or electronic lines projected from equipment on the ground.

Instrumentation in the cockpit at that time amounted to about a dozen flight instruments, including an **Artificial Horizon** (which shows the pilot the orientation of the aircraft relative to Earth and helps the pilot keep aircraft in level flight).

An **Altimeter** (which is an altitude meter used to measure the height of the aircraft above the ground).

A **VOR** radio display which is a round face instrument with a needle that centers when the aircraft is on track (aligned with the selected radial); for example, if the needle moves to the right the pilot steers to the left, in order to stay on course.

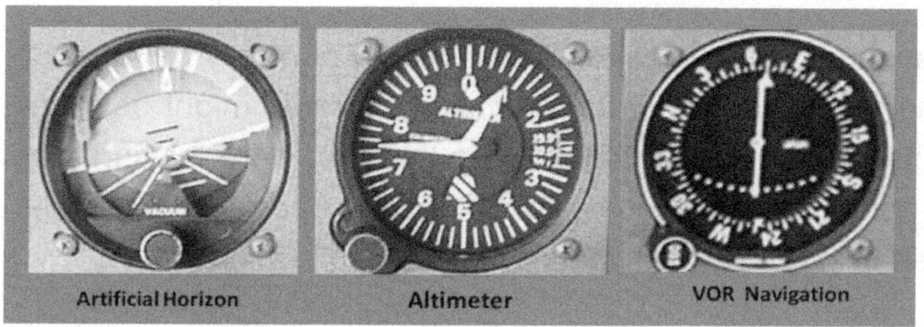

(Flight Instruments pictured here)

Aeronautical charts for VFR flight in 1976 were basically maps with the radials printed on them and included the corresponding frequencies that the pilot would use to tune the radio in order to track the electronic radial. The cockpit was lighted with red light to see controls and instruments inside the cockpit, as well as reading flight charts without affecting the pilot's night vision.

A small aircraft of that era would typically have four radios, two **Navigation VOR radios** as described above and two **Communication Radios** for voice transmission. Some also included a **Transponder** which is a radio device that transmits a signal (code number) to identify the aircraft on a radar screen.

The GPS navigation system in my car today is probably more sophisticated by comparison to the electronics in those old airplanes, nonetheless I can assure you that the pilot was understandably very busy back then with navigation, flying the airplane (keeping right side up), staying on course and talking to the **Air Traffic Controllers (ATC)**.

The voice communication radios would normally be tuned to separate frequencies, for example one radio might be tuned to the Departure Controller (area that you were leaving) and the other radio might be tuned to the Approach Controller (area of your next destination).

Aviation Voice Communication is very regimented and specific. There is rarely any kind of casual chit-chat between pilots and **ATC**. A radio transmission would typically start with the pilot identifying his aircraft for example, by saying EASTERN and the Flight Number, which would identify him as an Eastern Airlines flight.

Point being that Pilots would typically be monitoring such frequencies and listening to radio conversations (radio traffic) while in flight as I was that night. The following sequence is related from my memory of the strange event that occurred that evening:

- Out of the darkness I heard the voice of an Eastern Airline pilot make radio contact with ATC, requesting information about air traffic (other aircraft) at his flight level and in his flight path.
- After some hesitation, the ATC controller responded indicating that there was "no traffic identified on radar."
- A couple of minutes lapsed, and the pilot repeated the request, indicating they (pilots) had a visual on air traffic at

their flight level. (the tone of his voice was elevated, and he seemed irritated).

- ATC responded again, awkwardly "ah. . .ah. . . negative radar contact."
- The Eastern Captain contacted ATC again for the third time, the pilot described the traffic (object) with words like: **"huge" . . . "unconventional" . . . "hovering" . . . "stationary over the ocean" . . . "many lights on it" . . . "bright vivid multiple-colored lights" . . . "slowly rotating."**
- Quiet for a moment then the ATC controller hesitated and nervously responded "negative radar contact" It was obvious that there was something going on, but the ATC controller was not about to acknowledge radar contact, over the radio.
- At that point, another pilot's voice came over the radio, which identified as a small business jet. The second pilot said something spontaneous like "what Eastern is seeing so are we!"
- The second pilot's voice sounded as if he was very irritated when he advised that he had a visual on the same object. The second pilot did not specifically describe the object but indicated it was the same as the Eastern pilot described. No one wanted to articulate anymore details on the radio than necessary.

After that the radio traffic went silent! At that point, I was obviously very energized and kept looking out across the ocean trying to see it, but unfortunately it was too far out over the ocean for me to get a visual on it.

I was getting close to Jacksonville and contacted the control tower to get permission to land which was denied! ATC (Control Tower) instructed me to maintain holding (flying in a circle) for several minutes to accommodate what appeared to be two military fighter jets that were taking off from Jacksonville Naval Air Station. I was able to observe them as they came thundering off the runway and went booming out into the night sky over the ocean, engines lit with full afterburners in the general direction of the object the pilots reported.

The specifics of this event related herein are not as important as the three valuable lessons I learned that night:

1) There was for certain a huge unidentified, unconventional object hovering over the ocean that night which was reported by multiple pilots flying in two different jet aircraft viewing it from different perspective angles.
2) The Air Traffic Controllers refused to acknowledge the presence of the object and denied that they had radar contact.
3) The Air Traffic Controllers indirectly acknowledged the object by dispatching fighter jets to intercept it.

I flew back home to Daytona Beach that night without further incident. However, this event planted a seed deep in my consciousness and opened the door for a lifelong interest in the exploration of UFO Phenomena, which I believe is the mystery of the ages.

Formal Investigation Training

After graduation from Embry Riddle Aeronautical University, I worked during my professional life in the financial sector. Throughout the 1990s while working for a commercial bank as a Senior Trust Department Officer, I was afforded a lot of time off and disposable income to explore my interest in UFO research.

I belonged to a UFO organization known as MUFON which we'll talk a lot more about in chapter 5. I went through their training program to become a Certified Field Investigator and eventually was appointed as Chief Investigator for both Virginia and North Carolina. [3]

I also attended courses and trained to be registered as a Private Investigator in Virginia. (Virginia Department of Criminal Justice Services) I worked part-time for a private investigation firm based in Charlottesville and learned a lot about report writing, mobile surveillance, interviewing and interrogating witnesses.

Investigating Cases

As fate would have it, I found myself being pulled into one case after another. During those two decades, I interviewed over 30 abductees and was able to discuss with many of them their experiences in great detail.

I continued to grow and gain knowledge while investigating UFO cases. It is the little similar details that seem to surface from one story to the next that finally make you realize there is some resonant truth in the recurring details. Resounding patterns seemed to be developing through analysis of many unassociated cases and by comparing similar events that were described by people who were not associated with each other.

Strange Surveillance

I was investigating a case at a closed Government base in Virginia that we know now was an NSA facility while my wife was home alone with our dog. Our house was located in a secure gated community around Lake Monticello (near Charlottesville). No one could get into the neighborhood without police access or an invitation from a resident.

Our yard had no grass because the builders cut down just enough trees to make room for the house. The yards were typically wooded, and mulch was placed directly around the house and small back yard. Our dog's name was Wolfe. He was a big Husky/Malamute, weighing over 90 pounds. He had a leash run line behind the house that was attached to the back deck. We would let him out of the house from a sliding glass door to the back deck and he could walk around about a 50-foot radius from the back deck which was surrounded by thick woods.

My wife told me later, that while we were driving to the base, our dog (Wolfe) seemed to be agitated and was pawing at the sliding glass door to the back deck. By the time she got to the door he was extremely excited and appeared to be looking at something in the woods. She let him out by opening the door and hooking the leash run line to his collar. When she let him loose, he immediately ran to the end of his leash, pulling up, standing on his back legs and barking at someone who was in the woods. My wife noticed a man with a walkie-talkie at some distance in the woods who was standing there and looking directly at her, watching them.

One of the Investigators (who was a police officer) was riding with me when we returned to my home later that afternoon. When we heard my wife's story about the man in the woods, we decided to take a look around the community by driving around the lake. There was only one restaurant in the community inside the gate near the golf course. As we drove through the parking lot, we observed an unattended dark blue van in the parking lot. We actually got out and walked around the vehicle. There were windows in the back doors that were very heavily tinted but what I could see inside was typical of a surveillance vehicle.

There was a sticker on the front windshield "Department of Defense - Registered Vehicle - ANDREWS AFB" which I photographed.

We also observed a guest pass in the window which would indicate the vehicle was not from the community and they would have had to obtain access from the local police or with an invitation from a resident.

Later that evening we learned that several of the people involved with that investigation complained that their computers had been attacked with a virus that disabled them.

Pre-Digital Age

I am sure in that time period (1990s) some surveillance was maintained on UFO investigators in person, by essentially following them around to observe their activities and in some cases intercepting their mail from the postal service.

It should also be noted that this was in an era before the development of web-based UFO online reporting systems. UFO Investigation Reports were not on the Internet. For that matter, the Internet was still developing, and corporations were just starting to use email for communication in the mid to late 1990s.

UFO sighting reports were typed on paper forms and case reports were submitted in manila envelopes that were sent through the US mail (USPS). Photographs were taken with 35mm film cameras, and the film was sent to the local drug store to be developed and printed on photo paper. It should be noted that many older reports denote frustrated investigators that went to the drug store to find their film (negatives) and photo prints had been mysteriously picked up before they arrived.

The digital age was in its infancy. I remember having one of the first SONY Mavica digital cameras that used a floppy disc to record digital images.

Digital Age

Today information moves at the speed of light, with the Internet vast information is instantaneously available to the Investigator, not to mention that billions of people all over the world are walking around with the latest cellular technology and smart phones that incorporate hi-definition video cameras.

UFO case reports are posted on Internet sites that are web-based, that can include digital photographs, video uploads and documents all to be included in the case report databases.

Most of the clandestine surveillance which was done in the past by government agents is now electronically facilitated by intercepting email, utilizing GPS features in smartphones and through analysis of website visits which would obviously minimize the need to physically follow anyone around.

Government Service - Law Enforcement

During the 1990s most of us were concerned and almost paranoid about the government. Ufology was full of stories about Men-in-Black and government Bogie-Men sneaking around behind UFO witnesses and investigators. On more than one occasion I was worried about some of the witnesses and even my own personal safety.

Fortunately, I had the opportunity to serve in several law enforcement capacities including the Federal government and came to understand better how things work behind the scenes in law enforcement and government agencies.

Sheriff's Office reserve Deputy

I was doing public service as a Reserve Law Enforcement Officer in Virginia with the County Sheriff's office. In 2001 after graduating from the Police Academy, I became a Certified Law Enforcement Officer in Virginia in the aftermath of the 9-11 attacks. This was mostly about volunteer community service on weekends and evenings, and I really enjoyed it. [5] [6]

National Crisis 911

During this time Congress had mandated extremely high levels of security for USA Airports and the US Department of Homeland Security (DHS) was created. Like many people, I felt strongly about serving my country and temporarily left my banking career to work in the airport at Charlottesville,

Virginia serving in a dual role as a County Deputy and Special Deputy for the US Marshall Service.

US Department of Homeland Security (DHS)

I was offered a job to work with the US Department of Homeland Security as a Federal Liaison for TSA Stakeholders. In addition to my Corporate Management Experience and Airport Law Enforcement experience, I was a Licensed Commercial Pilot with a bachelor's degree in Aeronautics that obviously helped me get the position.

I served from 2003 to 2005 in the airports in Central Virginia. with TSA, I worked directly with the various law enforcement agencies concerning aviation security issues, including federal regulation and policy for airlines, airport authorities and law enforcement agencies including; FBI, ATF, ICE, State Police, Bomb Squads as well as the Airport Police, City Police, County Police and Sheriffs Offices.

During that tour of duty in a time of national crisis, **which had nothing to do with UFOs,** I learned a great deal about the inner workings of the various types of government agencies, which would later come in very handy in dealing with federal agencies and law enforcement.

END CHAPTER

Chapter 3

Government Cover Up - How it all Started!

As indicated, anything to do with UFOs, Aliens and Extraterrestrials is the most highly classified and carefully guarded secret in the U.S. Government. Let's take a look at some history to understand how this all started and what evolved.

Short Version

The short story is that during the late1930s and 1940s some of the governments around the world lead by the United States and Germany recovered crashed UFOs and took the wreckage to secret military locations. The crashes at Roswell and dozens of other locations supplied plenty of UFO wreckage and even extraterrestrial bodies for study. They used back engineering techniques and conducted secret research which originated the development of incredible technology.

Many believe the Nazi scientists surpassed the Americans in the development of advanced aerospace technology. Some speculate that a group of Nazi scientists during 1940s may have actually worked with military industrial and multinational technological corporations.

Operation Paper Clip

At the end of World War II, the American military in occupation of Nazi Germany took over the Nazi research facilities in Germany. They gathered up many of the Nazi scientists and confronted them with an offer to be repatriated to the United Stated and work in what would be NASA as well as other research facilities. This was officially known as Operation Paper Clip. During the late 1940s, 1950s and 1960s the United States supported numerous back engineering projects and accelerated technological research using these individuals.

Truman Administration and UFOs

Harry Truman became President of the United States upon the death of Franklin D. Roosevelt in 1945. He had been Vice President of the United States for only 82 days when he succeeded to the Presidency. During his first year in office, Truman approved the use of atomic bombs on Hiroshima and Nagasaki which caused the surrender of Japan and led to the end of World War II.

Atomic Testing

It was the testing of the atomic bombs and the destruction of the Japanese cities that many believe attracted the attention of those from out of town (Extraterrestrials from off planet).

As indicated UFOs have been visiting this planet for a long time as many historical references have been validated. However, with the onset of testing of atomic weapons the activity of reported UFO activity dramatically increased. Some believe atomic weapons may somehow disrupt the fabric of space in some form of Inter-dimensional atomic chain reaction off planet which may cause a threat to the Galactic Community.

More than a few UFO crashes occurred on American soil. (Listed at the end of this chapter) Research indicates that powerful anti-aircraft radar that was used in the early 1940s during World War II Era caused the anti-gravity propulsion systems in the saucer crafts to malfunction and result in UFO crashes.

James Forrestal and Majestic Twelve

In the late 1940s James Forrestal was Secretary of the Navy and managed UFO Projects for the Truman Administration. In 1947, President Truman appointed James Forrestal to be the first US Secretary of Defense, several weeks after the Roswell UFO crashes.

The day after the Roswell crash New Mexico Senator Dennis Chavez had a private meeting with President Truman. After talking with Truman, he personally called the media outlets in New Mexico and threatened to pull their broadcasting licenses if they continued to report the crashed UFO story.

Majestic Twelve (MJ-12)

In the wake of the Roswell Recovery Operations during meetings in the White House Oval Office an undertaking called Majestic Twelve (MJ-12) was developing and became operational to oversee the top-secret research dealing with the UFO issues while acting as an intelligence gathering agency.

MJ-12 ultimately controlled the reverse engineering programs in the 1950s and 1960s. Shortly after these Oval

Office meetings, President Truman issued a Memorandum instructing Forrestal to begin funding the MJ-12 group.

During my research I was surprised to learn that MJ-12 included members of the Council on Foreign Relations. There were 12 members of this group made up of the top officers of government, the military, and directors of the Council on Foreign Relations.

In early 1948 another UFO crash occurred in Aztec, New Mexico. Forrestal was involved in the UFO crash retrieval. These retrieval operations involved a number of alien bodies and wreckage of Extraterrestrial vehicles.

Privately his aides started noticing that Forrestal was developing some nervous mannerisms. This could have been because he was becoming more uncomfortable with the secrecy in hiding all this from Congress (and the public). There is no doubt that he could have been deeply disturbed by some of the things he was exposed to during crash retrieval.

He conveyed some of his concerns about the covert operations to Truman and wanted to brief some of Truman's political adversaries about the UFO situation which caused Truman not to trust him. There is intelligence indicating that at one point Forrestal may have threatened to talk to the media and press.

Forrestal Terminated

In any event, President Truman started to ease him out of office in early 1949 because of his quote "alleged declining mental health". He was reportedly shattered when Truman abruptly asked for his resignation. He was diagnosed with "severe depression" and ended up at Bethesda Naval Hospital in Maryland, where he fell to his death after going out of a window in his hospital room on the 16th floor.

Eisenhower Administration

1953 was Eisenhower's first year in office. There were 10 more alien crashed disks that were recovered along with 26 dead and 4 living aliens ETs.

Early in 1953 Eisenhower turned to a member of the Council on Foreign Relations, Nelson Rockefeller, for help with the alien dilemma. Rockefeller began coordinating the secret structure for control of the UFO situation by using MJ-12 and ultimately taking control out of the hands of Congress and the President.

During the Eisenhower Administration, the US government began to utilize technology corporations and aerospace firms to further develop the research.

In December 1959 the Inspector General of the Air Force issued the following Operations and Training Order: "Unidentified Flying Objects, sometimes treated lightly by the press and referred to as 'Flying Saucers', must be rapidly and accurately identified as serious Air Force business.

US Army Colonel Philip Corso

Colonel Philip Corso was an Army Intelligence officer during the Eisenhower administration. In a book written by Colonel Corso he explains how he was assigned to a secret government program in the Pentagon that provided material recovered from Extraterrestrial spacecraft crash wreckage to private industry to be used in reverse engineering programs. Colonel Corso outlined in the book many of the specific companies that the Pentagon shared the crash wreckage material with and how they developed technology from it.

For example, he named Bell Laboratories (now AT&T) who worked on fiber optic technology from tubing that was first recovered from wreckage. DuPont Laboratories developed Kevlar from material given to them by the Pentagon. IBM began developing the first micro-chip technology from recovered microcircuits just to name a few.

Remember back in the late 1950s electronic technology was carbon based (think of your grandparent's old radios with the big tubes inside) and then overnight by the 1960s the transistor age evolved, which is Silicon Technology.

These clandestine alliances were made through MJ-12 and tax money was funneled into classified back engineering projects. Deals were made to allow the companies to keep patent rights and make unbelievable profits as long as the technology was made available for government and military applications.

If you don’t believe anything in Colonel Corso’s book, just look at the companies he talked about from back in the late 1950s and you will see these corporations are now the multi-billion-dollar blue chip
corporations of today. [9]

President Eisenhower initially trusted Rockefeller. However, looking back, empowering Nelson Rockefeller was probably the biggest mistake Eisenhower made for the future of the United States and almost certainly all of humanity.

Beware of the Military Industrial Complex!

By the end of President Eisenhower’s administration, the Federal government's control in this research was diminishing thanks to Rockefeller and MJ12. Control was predominately in the hands of private multi-national contract operatives, global corporate technology and aerospace firms which President Eisenhower referred to as Military Industrial Complex. When he left office, he warned us about their misplaced power and control in that famous speech “Beware of the Military Industrial Complex”.

Eisenhower threatened to invade Area 51

There is an incredible video that was produced by Richard Dolan interviewing an elderly gentleman who worked for the CIA and was part of Project Blue Book (an Air Force UFO investigation operation).

As part of his work in Project Blue Book he described how they met with witnesses and tried to discredit them, smiling and saying something about Swamp Gas theories. In the video he describes how he was dispatched to the oval office with his boss to meet with Eisenhower.

He claimed to have a “Q Clearance” which is the highest Security Clearance in the United Stated that is well above TOP SECRET. He claimed to have been with President Eisenhower and Richard Nixon in several meetings about Area 51 the Groom Lake Testing Facility and at S4.

Eisenhower was concerned that the operations at Area 51 and S4 were being kept from him and Congress. He describes how Eisenhower instructed him and his boss to go to Area 51 and deliver a message. To tell them they had a week to brief the president on the operations at the facility or Eisenhower threatened to come to Colorado, take control of the first army and invade the base.

He describes in detail how he and his boss went to Area 51 and then were escorted to S4 where they were shown alien bodies and saucer crafts that used reverse gravitational propulsion. Eisenhower let them use his Lockheed Electra (which was essentially the first Airforce One). They returned to the White House and reported to Eisenhower who was very worried about how the Black Projects had grown so out of control.

The man is elderly and in bad health but very believable and worth watching. A link to Richard Dolans video can be found in the references at the end of the book. [10]

JFK and UFOs

President Kennedy was a Naval Intelligence officer during his military career. Research indicates that his father Joseph Kennedy Sr, who was a very wealthy powerful businessman and American politician contacted his friend James Forrestal and asked him to take Jack (JFK) under his wing and help to groom him for a possible political career.

A long-term alliance developed between JFK and Forestall. JFK was in Germany in the aftermath of the war and research indicates that by then he had a close relationship with James Forestall who was Secretary of the Navy at that time.

Form my experience in the early days, Naval Intelligence was a main player in the UFO government research. It is very likely that Kennedy knew firsthand about the Germans Nazi UFO research.

Kennedy served in the House of Representatives from 1947 to 1953. Then later as a US Senator from 1953 until his election in 1960 to President of the United States.

Because of his Naval Intelligence background and the fact that he had been privy to the UFO research by the Nazis, he was one of the US Congressmen trusted with the Roswell secrets. Some sources indicate that thanks to James Forrestal, he may have been involved in some capacity during the recovery operation.

When Kennedy succeeded Eisenhower as president he was equally concerned about the Military Industrial Complex and Black Projects that had gotten out of control.

Historical accounts indicate that JFK and Eisenhower met secretly in the White House on several occasions for dinner and cocktails where they were able to candidly discuss their concerns.

Kennedy, being prompted by Eisenhower, was obviously genuinely concerned about the operations at Area 51 and S4. History tells us that his inquiries were stonewalled in the same way that Eisenhower had been treated.

The President of United States is the Commander in Chief of all military services and bases, he can visit any military base at any time desired. It is interesting to note that Area 51 and S4 are <u>not</u> U.S. military bases, but rather secret facilities controlled primarily by the CIA (Central Intelligence Agency)

Note: The CIA has been closely involved in collection and suppression of UFO information. "Witnesses to the UFO phenomena have been bribed, coerced and threatened by the CIA in many documented cases.

It is pretty well acknowledged the President Kennedy and President Nixon had been to the MacDill Airforce Base in Florida where alien bodies and crafts were apparently stored. I was at MacDill Airforce base in the late 1970s for high altitude flight training but didn't get to see any alien bodies or crafts.

After the Cuban Missile Crisis JFK had great concern that a UFO event might be misinterpreted as a missile strike and cause a nuclear war. He had apparently been communicating with Khrushchev in a consorted effort to share information about UFOs with the Soviet Union to prevent a potential nuclear disaster that might be triggered by a UFO.

His contempt for the CIA in his promise to break the CIA into a thousand pieces and the communication with USSR about UFOs probably caused them to terminate his life with regrets. The disaster at Dealey Plaza in Dallas was most likely orchestrated by the CIA.

Interestingly the bodies of both JFK and his mentor James Forrestal ended up being autopsied at the same place, at Bethesda Naval Hospital in Maryland.

COMETA - France

In 1999 a document was published in France entitled **"UFOs and Defense: What must we be prepared for?"** This ninety-page report is the result of an in-depth study of UFOs, especially addressing questions of National Defense. The study was carried out in France over several years at the Institute of Advanced Studies for National Defense (IHEDN) and through COMETA which translates to "Committee for in-depth studies and involved both government and military officers. [15]

The COMETA report officially recognizes Colonel Philip Corso and concludes that his testimony (book) is most likely fully accurate and conveys the real situation in the United States Government.

The report explores motivations of Extraterrestrial visitors such as protection of planet earth against the dangers of nuclear war, suggested by frequent events of UFOs flying over nuclear missile sites.

Relationships with Extraterrestrial Civilizations

The COMETA report examines the official behavior of different nations and focuses on the probability that secret, privileged contacts have been established with the United States. By contacts they mean direct relationships with off planet Extraterrestrial Civilizations.

Also suggesting that since 1947 and the Roswell event a policy of ever-increasing secrecy seems to have been applied to protect US Military technological research and back engineering projects using recovered UFO wreckage and has been controlled exclusively by the Military Industrial Complex.

My point is to help the reader understand the power of the secrecy and control surrounding the Military Industrial Complex. Disclosure is not so much about UFOs, Aliens and Extraterrestrials. It is about removing of the secrecy surrounding the Military Industrial Complex and the control they have over the American Government and Global Economy which is by far more dangerous than any threat from Extraterrestrial invasion.

Deep State

The Military Industrial Complex operates as part of the Deep State. The Deep State is essentially defined as the influence of career civil servants over elected officials, it's not only the intelligence agencies but the career bureaucrats of government who sit in powerful positions, who don't leave when Presidents do, who see presidents come and go. They view Presidents and other Elected Officials as temporary employees. They have long been concerned about Incumbent Presidents disclosing secret projects with military and private defense contractors.

They influence policy and can operate in opposition to the agenda of the elected officials by obstructing, resisting and subverting their policies and directives, thus constituting a hidden government within the legitimately elected government.

The Deep State consists of networks of power operating independently of the nation's elected political leadership in pursuit of their own agenda and goals. The Deep State operates through many front **organizations** and associations with multinational corporations in a variety of **industries**.

These organizations orchestrate political and financial events as well exerting control on both national and international events. These interconnected industries wield immense power and control.

INDUSTRIES

MEDIA
- Main Stream Media
- Social Media
- Hollywood
- Control News and Social Engineering

FOOD and DRUG
- Agricultural
- Pharmaceutical
- Control Food Supplies and Medical Research

MILITARY INDUSTRIAL COMPLEX
- Defense Contractors
- US Military
- Intelligence Agencies
- Global Police and Enforcement
- Control Global Conflict and Thrive on Wars

PETROLEUM
- Petrodollar
- Global Price Fixing
- Control the World's Energy Supply

CENTRAL BANKING
- Federal Reserve
- Fiat Currency
- Control the World's Money Supply

TECHNOLOGY
- Silicon Valley
- Human Profiling
- Digital Tracking
- Control with Artificial Intelligence

Council on Foreign Relations

Probably the most powerful of these organizations is the Council on Foreign Relations which is a United States nonprofit organization specializing in US foreign policy and International Affairs. It is headquartered in New York with an additional office in Washington, DC. This is the group that many believe are actually operating the secret government of the United States. The members are too many to name, but they include US presidents, senior politicians, more than a dozen Secretaries of State, CIA Directors, bankers, lawyers, professional media figures, journalist and Hollywood actors. Virtually every important politician in our government as well as CEOs of giant corporations, banks and media are members of the Council on Foreign relations, which explains why there's a news blackout in the mainstream media on any coverage of their meetings. We know from the Eisenhower Administration that approximately half of Majestic 12 members were chosen from this organization.

It is believed that every President after Ronald Reagan has served the Deep State which grew stronger with each succeeding President as they gained control over everything behind the scenes. if you don't think so just go and listen to George Bush senior in his famous speech about The New World Order.

Very High Technology

Fast forward 50 years and realize that these technology corporations and aerospace firms operate outside the control of any government, Including the United States (that originally funded them and provided alien technology for their research).

Advance Technology associated with free energy production, anti-gravity transportation (vehicles) and medical advances are not available to the public whose tax dollars paid for the research.

In secret, they have developed advanced technology that is 100 years ahead of anything we currently have in the Public Domain and have made billions of dollars for their own purposes.

Extraterrestrial Vehicles and Terrestrial Technology

There are clearly Alien and Extraterrestrial vehicles that visit this planet piloted by different species from different places in the galaxy, from different dimensions and different times. Also, there is terrestrial technology that is tested and utilized by the Aerospace and the Military Industrial Contractors. Many of the triangle shaped objects are thought to be in this category.

UFO Crash Wreckage

This is much vaster than most people would expect, even those of us who have spent many years involved in this research. To put this in perspective we acquired the following list of UFO crashes that occurred where wreckage and alien bodies have been recovered.

Date - **Location** - ET Bodies

July 4, 1947 **- Roswell, New Mexico** - 4 Bodies

Oct 1947 - **Paradise Valley Arizona** - 4 bodies

Feb 1948 - **Aztec, New Mexico** - 12 Bodies

July 1948 - **Mexico, South of Laredo** - 1 Body

1949 - **Roswell, New Mexico** -1 ET Living

1952 - **Spitsbergen, Norway** - 2 Bodies

Aug 1952 - **Ely, Nevada** - 16 Bodies

Sept 1950 - **Albuquerque, New Mexico** - 3 Bodies

April 1953 - **South West, Arizona** - No Bodies

May 1953 - **Kingman, Arizona** - 1 Body
June 1953 **- Laredo, Texas** - 4 Bodies
July 1953 - **Johannesburg, South Africa** - 5 Bodies
Oct 1953 - **Dutton, Montana** - 4 Bodies
May 1955 - **Brighton, England** - 4 Bodies
July 1957 - **Carlsbad, New Mexico** - 4 Bodies
1961 - **Timmensdorfer, Germany** - 12 Bodies
June 1962 - **Holloman, AFB, New Mexico** - 2 Bodies
Nov 1964 - **Ft. Riley, Kansas** - 9 Bodies
Dec 1965 - **Kecksburg Pa** -No Bodies
Oct 1966 - **North West, Arizona** - 1 Body
July 1972 - **Morocco, Sahara Desert** - 3 Bodies
July 1973 - **North West, Arizona** - 5 Bodies
Aug 1974 - **Chihuahua, Mexico** - 0 Bodies
May 1976 - **Australian Desert** - 4 Bodies
June 1977 - **North West, Arizona** - 5 Bodies
April 1977 - **South West, Ohio** -11 Bodies
Aug 1977 - **Tabasco, Mexico** - 2 Bodies
May 1978 - **Bolivia** - No Bodies
Nov 1988 - **Afghanistan** - 7 Bodies
May 1989 - **South Africa** - 2 ET Living
June 1989 - **South Africa** - 2 ET Living
July 1989 - **Siberia** - 9 ET Living
Sept 1990 - **Megas Platonos, Greece** - No Bodies
Nov 1992 - **Long Island, New York** - No Bodies

If even half of this list is accurate and I am reasonably sure it is, you can see these guys have been remarkably busy recovering wreckage, doing extensive research and making billions of dollars while developing technology that is literally out of this world.

Encounters with Extraterrestrials on the Moon.

This is an amazing story that will push most people to the edge of their comfort level. It is about close encounters with Extraterrestrials during the Apollo 11 Space Mission and first walk on the Moon. I am quite sure this is true and accurately relayed herein as I have researched it thoroughly. However, I am convinced it will challenge most people's perception of reality.

HAM Radio Operators on Earth

The radio communication between NASA Houston Space Center and the Apollo 11 Crew was intercepted by numerous amateur HAM radio operators on Earth. This was obviously a time before sophisticated radio encryption equipment which would be used today would prevent unauthorized radio operators form receiving such transmissions.

Confirmation of Apollo 11 Lunar Event

Several years ago, I published an article in a local magazine in Charlottesville about this Apollo 11 event. I spoke with someone I knew well in Charlottesville telling him about the article that I was going to publish in the next edition,

My friend told me about a special project he had worked on at the McCormick Observatory in the summer of 1969. This individual is someone I know to be highly credible, well-educated to the level of PHD and has asked to remain anonymous.

McCormick Observatory, Charlottesville Virginia

At that time the telescope at the McCormick Observatory (pictured here) in Charlottesville was State of the Art highly sophisticated and one of the two most powerful telescopes in the world.

The purpose of his project was to explore the surface of the Moon in order to identify potential landing spots for Apollo 11 Space Mission. He articulated to me that while exploring the surface of the moon with the telescope he personally viewed UFO traffic and lights on and over the surface of the moon, just prior to the Apollo 11 Space Mission Landing.

In 1979 Maurice Chatelain was the Chief of the NASA Communications Systems. He made a very public statement that: "All Apollo and Gemini flights were followed both at a distance and sometimes also quite closely by space vehicles of Extraterrestrial origin. Every time it occurred, the astronauts informed Mission Control, who then ordered absolute silence."

Mr. Chatelain also acknowledged in his public statement in 1979 that: "Neil Armstrong had indeed reported seeing two UFOs on the rim of a crater. The encounter was common knowledge in NASA, but nobody has talked about it publicly until now."

Apollo 11

Apollo 11 was the first space mission to land on the Moon. Neil Armstrong, Michael Collins and Buzz Aldrin manned this Lunar Flight Mission. Collins piloted the Command Module around the Moon (remaining in orbit) while Armstrong and Aldrin descended in the Lunar Landing Module (LEM) touching down on the surface of the Moon at 4:17 P.M. at a location on the Moon known as "Sea of Tranquility" on July 20, 1969.

Neil Armstrong relayed messages to Houston Mission Control that two large mysterious objects were watching them after landing near the lunar landing module. But his message was never heard by the public because NASA censored it.

Christopher Kraft was director of NASA Houston Space Center during the Apollo Projects. He made some interesting comments about this Apollo 11 event when he left employment at NASA. His comments about the transcribed excerpts from radio transmissions have been corroborated by hundreds of amateur radio operators who had tuned in their stations to the same frequency that the astronauts communicated with NASA Houston Space Center (Houston Control).

Ham radio operators who had their own VHF receiving facilities (the guys with the massive antennas in their back yards) picked up the following amazing conversations. At the same time the live television broadcast was interrupted for several minutes due to a supposed "overheated camera", but the transcribed radio transmissions (below) were received loud and clear by over 700 ham radio operators in the USA, while an estimated 600 million people around the world (myself included) were watching and listening to the official version on TV.

Buzz Aldrin and Neil Armstrong were walking around on the Moon in view of the camera some distance from the Lunar Landing Module (LEM) when Armstrong suddenly clutched Aldrin's arm excitedly and exclaimed (as the camera and sound went off during the public live TV broadcast):

- **Apollo 11 (Moon):** (Neil Armstrong) What was it? What the hell was it? That's all I want to know!"
- **Mission Control: (**Christopher Kraft-Houston Space Center) what's there? ... (garble) ... Mission Control calling Apollo 11.
- **Apollo 11 (Moon):** These babies were huge, sir! ... Enormous! ... Oh, God! You wouldn't believe it! I'm telling you there are other spacecraft out there ... lined up on the far side of the crater edge! They're on the Moon watching us!

- **Apollo 11 (Moon):** Those are giant things. No, no, no this is not an optical illusion. No one is going to believe this!
- **Mission Control:** What ... What ... What? What the hell is happening? What's wrong with you?
- **Apollo 11 (Moon):** They're here under the surface.
- **Mission Control:** What's there? (muffled noise) transmission interrupted . . . Mission Control calling Apollo 11.
- **Apollo 11 (Moon):** We saw some visitors. They were here for a while observing the instruments.
- **Mission Control:** Repeat your last information!
- **Apollo 11 (Moon):** I say that there were other spaceships. They're lined up in the other side of the crater!
- **Mission Control:** Repeat, repeat!
- **Apollo 11 (Moon):** My hands are shaking so badly I can't do anything . . . Film it ... God, if these damned cameras have picked up anything . . . what then?
- **Mission Control:** Have you picked up anything?
- **Apollo 11 (Moon):** Three shots of the saucers or whatever they were that were ruining the film.
- **Mission Control:** Control here . . . are you on your way? What is the uproar with the UFOs over?
- **Apollo 11 (Moon):** They've landed here. There they are and they're watching us.
- **Mission Control:** The mirrors, the mirrors - have you set them up?

- **Apollo 11 (Moon):** Yes, they're in the right place. But whoever made those spaceships surely can come tomorrow and remove them. Over and out.

During a NASA symposium several years later, a professor was interviewing Neil Armstrong about this event. Note: Buzz Aldrin apparently took color movie film of the UFOs from inside the module and continued filming them after he and Armstrong went outside.

- **Professor:** What REALLY happened out there with Apollo 11?
- **Armstrong:** It was incredible, of course we had always known there was a possibility, the fact is, we were warned off! There was never any question then of a space station or a moon city.
- **Professor:** How do you mean "warned off"?
- **Armstrong:** I can't go into details, except to say that their ships were far superior to ours both in size and technology. Boy were they big! And menacing! No, there is no question of a space station.
- **Professor:** But NASA had other missions after Apollo 11?
- **Armstrong:** Naturally, NASA was committed at that time, and couldn't risk panic on Earth.

Armstrong confirmed that the story was true but refused to go into further detail beyond admitting that "the CIA was behind the cover-up."

In a published article in April 1995, Maurice Chatelain who was former NASA Director of Communications dropped the bombshell revelation stating that "the Apollo Moon Missions found several mysterious geometric structures of unnatural origin on the Dark side of Moon".

Clearly the secrecy gap surrounding government activities is beginning to narrow due in part to Internet communication technology, general increase in knowledge and an elevation in human consciousness.

We will talk more about current news events relating to disclosure that are developing later in chapter 7

END CHAPTER

Chapter 4
Covert Government Operations

It has been said that the secrecy surrounding the UFO/UAP phenomena and the covert back engineering programs have been classified higher than the research for the development of the Atomic Bomb.

Project Blue Book

Project Blue Book was one of a series of systematic studies of UFOs conducted by the United States Air Force. It started in 1952, and it was the third study of its kind. [11]

The first two were **Project Sign (1947)** and **Project Grudge (1949)**. We have the benefit of 70 years of history and the stories of investigators associated with Project Blue

Book who interviewed UFO witnesses, confiscated their photos and did the best they could to discredit and intimidate them. It seems that government or contract operatives in keeping with the need for UFO secrecy were conducting covert surveillance on witnesses, investigators, and experiencers in order to gauge the extent to which the public was aware of UFO/UAP phenomena and the existence of Extraterrestrials.

Project Blue Book ended in December 1969.

UFO Organizations

In addition to MUFON (Mutual UFO Network) some of the other organizations that study UFO phenomena are:

- **NICAP** (National Investigation Committee on Aerial Phenomena).
- **NUFORC** (National UFO Reporting Center)
- **CAUS** (Citizens Against UFO Secrecy).
- **GSW** (Ground Saucer Watch),
- **CUFOS** (the Center for UFO Studies)
- **APRO** (Aerial Phenomena Research Organization), an Arizona nonprofit scientific and educational organization, founded in 1952

It is interesting to note that some of these Grass Roots UFO organizations started to solicit reports for UFO sightings in 1970. Is it a coincidence that UFO organizations like **MUFON** and **NUFORC** started at almost the exact same time that Project Blue Book ended?

I have often wondered if this was a new way for government agencies to use newfound UFO Organizations to monitor the public awareness about UFO activity without the extreme expense associated with covert surveillance of witnesses and investigators.

Not to mention the plausible deniability aspect which allows government officials in a formal chain of command to deny knowledge or responsibility for any actions committed and that private organizations are not subject to FOIA (Freedom of Information Act) requests.

The most important element of any UFO organization is the Data Base. In the early days as indicated, UFO Investigation reports were typed on paper and case reports were submitted in manila envelopes that were sent through the US mail. These reports were not generally available to the public.

UFO Reports and Data Bases

With the development of Internet data bases and automated reporting sites on the Internet, these computerized case management systems like the one used by MUFON and NUFORC developed. Individuals that have a sighting or experience can go online and submit a report.

These report websites have evolved to be quite complex and allow the individual to report in detail the experience by answering questions and filling in blanks on the digital report form. Photographic images, audio recordings and video can be uploaded as well. These on-line reports feed into a database that can be sorted for a variety of reporting applications. As the database grows over the years the data becomes more useful and valuable.

Databases, who has the keys to the back door?

In the case of MUFON limited access to the data is also made available to members of the organization. The design and maintenance for the servers of such an extensive computer system would certainly be very costly. During my time with MUFON I was never able to understand how this massive system was financed. There were no records of payments for the database, and no one could explain how it was paid for.

Some of us who have had unlimited access to these reporting system databases speculate that they were likely designed by very sophisticated technology corporations like IBM and government contractors. On the surface, these databases appear very legitimate to the user, as an individual submitting a report or the investigator using the system. But hidden from public view are complex back doors and control systems which allow those who design these systems unlimited access to utilize and sort data as well as conceal extremely sensitive UFO reports.

Irrespective of what is being done behind the scenes, many of the field investigators and grassroots members still benefit greatly from the experiences that they have investigating cases from the Case Management Systems.

GPS Coordinates for Investigative Reports

Another thing which seems strange is most UFO data bases specifically MUFON, now require GPS coordinates for all Investigative Reports. This was something that we started during the BAASS MUFON Sip Project. We were required to include GPS Coordinates (latitude and longitude) with exact times and dates of the events on all reports submitted to BAASS.

Regressive Satellite Imaging

No one at BAASS ever said so but I kind of figured that a multi-billion-dollar Aerospace company like Bigelow might have access to some satellite technology where they could use regressive imaging (knowing the exact date, time and coordinates of an event, you can go back and review recorded images from satellites that orbit the earth and photograph the surface). To my knowledge no UFO organizations have access to regressive imaging satellite technology, but sophisticated technology corporations and government agencies do which makes an interesting case for questioning the requirement for GPS coordinates.

Going back to the idea of monitoring the public at large to know what the public is seeing, perceiving, photographing as well as who is communicating on any level with whatever extraterrestrial or inter-dimensional entity is unbelievably valuable information. Also, what is NOT seen or perceived by the public may also be valuable.

Operation Moon Dust

In the 1950s a secret project called "Operation Moon Dust" was initially responsible for retrieval of downed space objects including UFOs that operated out of a special unit at Fort Belvoir known as the 4602nd. The 4602nd became the 1006th Air Intelligence Service Squadron (AISS) in July 1957. In April 1960 the 1006th AISS became the 1127th USAF Activities Group Air Reserve Squadron,

Crash Recovery Retrieval Teams (Recon Units)

There are still Crash Recovery Retrieval Teams (Recon Units) that are always on stand-by to deploy to the scenes of such events. Most of these Recon Units still operate out of Fort Belvoir and may be a combination of military and or government contractors. Contractors can operate covertly with less accountability than the Military.

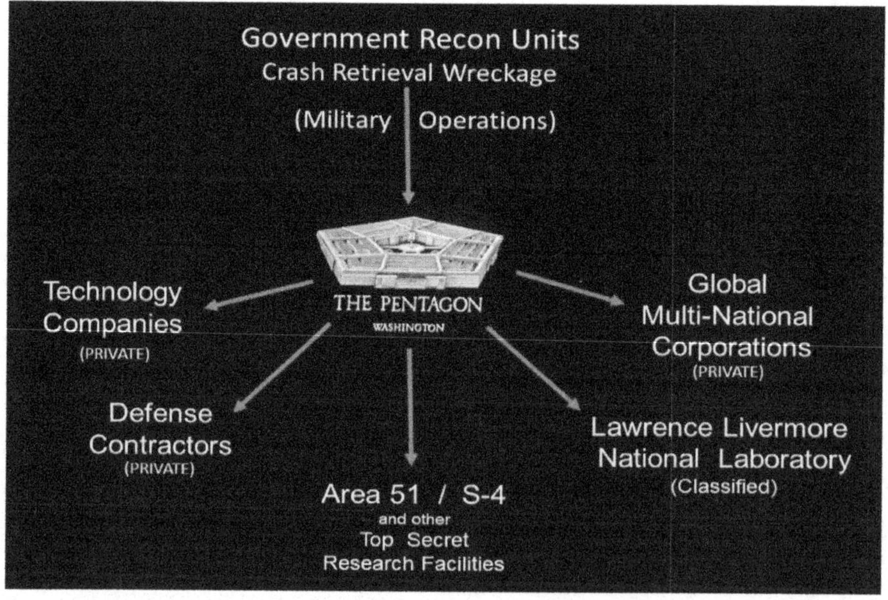

These Recon Units are not necessarily always involved with UFO / UAP events. They would also respond to a satellite crash, lost missile warhead accident, military aircraft crash including retrieval of a lost nuclear armament as well as any type of space debris coming down.

Let's explore a hypothetical CRASH situation!

The first order of business would be to set up a perimeter around the area where an event occurs by blocking streets, railroad lines and interstate highways leading to or around the area. The surrounding airspace would be immediately restricted to all air traffic (including drones).

POLICE LINE DO NOT CROSS

A cover story would be sent out to the media to explain the perimeter, like a train wreck, plane crash, quarantine for disease or the accidental release or spill of toxic substances.

Local police would likely be involved and would assist with perimeter control and access points. Police would not typically be involved in the retrieval process and would probably not know the true nature of the event (or what was actually going on inside the perimeter). However, the cover story would be serious enough to justify the use of force to maintain control of the perimeter.

Recon Units

The vehicles pictured here were traveling in a caravan, which included about 15 vehicles at an undisclosed location in Virginia. The photos which were taken from public view and show how a Recon / Recovery Unit might appear on the highway while in-route to an anomalous event.

These vehicles have high tech satellite microwave receivers and antennas mounted on top and all the vehicles. The vehicles were painted dark blue and appeared to be in brand new condition. Note that police light configurations (flashing Red and Blue) were mounted on all these vehicles.

Pictured here is what appears to be a mobile command center and they were all manned and operated by what appeared to be uniformed military officers. All vehicles had Federal Government Plates

In the event of a crash or a landing Recon Recovery teams would immediately deploy to the scene of an anomalous event.

Serious Nature of Crash Recovery Retrieval

In my first book we included this Warning to UFO Investigators which serves to indicate the serious nature of Crash Recovery Retrieval Operations.

These guys like the ones pictured on the previous page would immediately take control of the area and are authorized by Federal Law to arrest and detain indefinitely anyone who confronts them or breaches the security perimeter they set up. Don't even think about sneaking into an area already set up with a security perimeter!

What if you get there first?

Being in the wrong place at the right time, here are some things to consider:

- Do <u>not</u> ever cross a police line!
- Personal safety is most important!
- Crash debris could be radioactive, electromagnetic or discharge toxic substances. Note that some satellites have small nuclear reactors which could obviously leak after a crash landing.
- Do not walk or stand under a hovering UAP as the anti-gravity (electrogravitic) propulsion emits a type of electromagnetic energy, similar to microwave in the way that it will cause skin tissue damage.
- Do not touch or attempt to touch a UAP that has landed. [6]
- A UAP that is landed on the ground and still powered up can emit ultraviolet radiation from the surface of the craft that will cause burns to the skin like sunburn, also there may be an emission of radiation in the X-ray spectrum that can cause radiation sickness.
- Do not point any kind of high intensity flashlight or high-power spotlight at a hovering UAP! There are known cases where persons have been burned by bright light rays emanating from UFOs after doing so.
- Psychological effects produced by force fields that could induce a hypnotic state in the observer, including loss of consciousness, memory relapse, and submission to the UFO occupants.
- It goes without saying that any display of firearms or other weapons on your part could be construed as hostile and will most likely end disastrously.

- The military, police, or government agents that respond will <u>not</u> be nice to you! You can forget about any civil rights you think you have in a place like this, especially if the military or government security contractors are involved, in which case the rule of law will not apply.

- You could be treated as an enemy combatant with no right to legal representation including the right to remain silent. Not to mention you may not be allowed to contact family members.

- When you vacate the event site area, you may be stopped by police in the surrounding community. They will question you and will definitely search your vehicle, the less you say the better chances you have of getting out of there.

- Your Smart phone is your worst enemy so don't call your spouse or significant other to brag about what you might have just witnessed.

- Police and Federal agents have technology that they use to do surveillance on drug traffickers and other criminal activity (targets a specific area) that works very efficiently with smart phones. They can track your movement, listen to your phone conversations and in some cases listen to and see video of you when you think the phone is turned off. Same with your laptop or iPad if it has a modem connected to the Internet.

- If you think you are being monitored turning the smartphone off will not work. They can still covertly turn it on and listen to your calls and see your texts and email.

Take the battery out of the smart phone or leave it in the car or at the hotel.

Note: Lying to a federal agent is a federal offence punishable with imprisonment! Information contained herein is not intended to be legal advice, so it is best to consult an attorney for more on this as laws and restrictions vary from state to state.

NORAD

NORAD (North American Aerospace Defense Command) safeguards the airspaces of the United States and Canada by monitoring unknown and unauthorized air activity approaching and operating within our airspaces.

NORAD monitors inner space so carefully that a small piece of debris from a satellite is scrutinized and cataloged. There is nothing that enters the atmosphere of the earth that is not detected! There is nothing that materializes in the airspace above the earth that is not detected!

Therefore, any UFO/UAP event (sighting) that occurs is likely known and monitored by NORAD. Some are investigated and sometimes intercepted by military aircraft. This explains why so many sighting reports indicate that witnesses observe a UFO that is followed by Fighter Jets or

Black helicopters. If that is so, then the question arises as to why the government would be so interested in UFO sightings, witnesses and civilian investigations when if in fact these agencies know exactly what is happening? [12]

The Office of Global Access (OGA), a wing of the Central Intelligence Agency's Science and Technology Directorate, has played a central role since 2003 in orchestrating the collection of what could be alien spacecraft.

CIA has allegedly coordinated the secret recovery and storage of alleged crashed or landed UFOs. They have a 'system in place that can discern UFOs while they're still cloaked,' and that if the 'non-human' craft land, crash or are brought down to earth, special military units are sent to try to salvage the wreckage. Its operations also involve more conventional retrieval missions, such as stray nuclear weapons, downed satellites, or adversaries' technology. [17]

United States Space Force

The U.S. Congress authorized the creation of the United States Space Force, which was officially established on Dec. 20, 2019.

The National Defense Authorization Act ultimately facilitated the creation of the United States Space Force which has immense powers and authority not to mention weapons that are yet to be heard of in the public domain. The commanding General of the United States Space Force sits with the Joint Chiefs of Staff and is among the most powerful Generals.

Many believe the space force was created in part to ultimately legitimize the UFO research and bring it into the public domain as part of the long-term disclosure project. In the near future we're going to hear a lot more about the United States Space Force.

The PRESS and UFO Investigators

Until very recently the media and press were not your friends when conducting UFO research. As indicated, this has dramatically changed in the last year or so.

The first MUFON symposium I attended was in 1999 at a hotel complex in Washington DC. I recall being so excited about the many interesting people there from all over the world. Many writers, speakers, documentary producers, researchers, scientists, and world class investigators were in attendance. I would guess that there were at least 80 people there educated to the level of PHD.

I remember when we pulled up in the drive-through at the hotel, there was some guy out there walking around with a big tinfoil hat on his head acting kind of goofy. He was obviously positioned there to embarrass and discourage the individuals who were attending the UFO symposium.

My point in documenting this is after the symposium was over one of the Washington newspapers published an article about the UFO Symposium. Interestingly of all the professionals and PhDs that attended the symposium, you can guess whose picture was on the front page. Yes, the goofy guy with the tinfoil hat was on the front page of the newspaper and as I remember the article was not very complimentary, talking about the conspiracy theory people getting together at the conference center.

TV Shows and Documentaries

Generally speaking, the rules for producing television shows and documentaries in the United States about UFOs required that the producers interview at least one participant that blatantly debunks what has been presented. I am not sure there is any research that indicates any specific State or Federal laws pertaining to this, but generally the networks and the media companies in the USA all followed the same rules.

For example, if they produce a one-hour documentary about a UFO case, first the investigators present information, pictures and testimony about the event or case they investigated. Then at least 25% of the remaining time will be allocated to someone with an adversary view which usually comes at the end.

So, during the first part of a typical UFO television show the investigators and witnesses enthusiastically present the case and evidence, then during the last 15 minutes a guy wearing a bowtie with a Cheshire Cat Grin on his face comes on smugly claiming to have scientific credentials and basically discredits the witnesses characterizing them as conspiracy theory idiots and debunks everything the investigators said. A lot of us would like to think there is a special place in Hell for that Bowtie Bastard.

TV Studio Green Room

Often investigators participating in network television broadcasts were separated and isolated in a greenroom while being interviewed. They could communicate with the commentator through an earpiece so that they could hear the commentator's questions but could not see or hear what the other people on the show were talking about. Figuratively speaking it is pretty much like being blindfolded and then pushed off a cliff.

MEDIA outside of the USA

In all fairness I would like to say the rules about filming in Canada were much different. In 2012 and 2013, we flew to Toronto a number of times to work with **Discovery Channel Canada** on a documentary series about UFOs that was called "**Close Encounters**".

The producers created two seasons of “Close Encounters”. I was filmed and interviewed in several episodes of the First Season and the following year invited back to work on Season 2 (pictured here). We worked out of a studio in Toronto and it was really fun. The shows ultimately aired in Canada, Europe and the United States

As indicated, I have worked extensively with Discovery Channel filming here in the United States. I can honestly say that working with Discovery Channel Canada was a much better experience. The producers in Canada were considerably more open minded and objective about the subject material.

The rules in Canada are quite different, they apparently have no obligation to utilize the adversary witnesses or debunkers in any part the shows and documentaries.

In Mexico the subject of UFOs is much more mainstream and accepted by the media (and the general public). Mexican television newscasters seem to be much more open minded and objective about the subject and have broadcaste some incredible video and pictures to the public.

END CHAPTER

Chapter 5
DIA - BAASS - MUFON - ASSWAP

In 2008 **Bigelow Aerospace** was an American Space Technology Company and Defense Contractor based in Las Vegas, Nevada. [7]

BAASS (Bigelow Aerospace Advanced Space Studies) was a sister company to the Bigelow Aerospace Corporation and a research organization that focused on the identification, evaluation, and acquisition of novel and emerging future technologies worldwide specifically related to spacecraft. BAASS was headquartered in a secure underground complex at Bigelow Aerospace in Las Vegas.

MUFON (Mutual UFO Network) is a non-profit civilian organization that investigates cases of reported UFO sightings. MUFON was founded in 1970 and is the largest international UFO research organization in the world. Over a period of forty plus years MUFON has created a very complex data base comprised of UFO witness reports and investigations. These reports were submitted by witnesses from the general public and accumulated on a web based MUFON site known as Case Management System. [4]

Defense Intelligence Agency (DIA)

Behind the scenes in 2008 the Defense Intelligence Agency (DIA) was secretly preparing to fund several UFO/UAP research programs which were ultimately associated with MUFON through the Pentagon's **Advanced Aerospace Weapon Systems Applications Program (AAWSAP)**

ASSWAP was part of a program sponsored by then Senate Majority Leader Harry Reid at the urging of his Nevada billionaire friend Robert Bigelow to investigate UFOs. It was known as the **Advanced Aerospace Weapon Systems Applications Program (AAWSAP)** and was run by the DIA with a budget of $22 million over its five years of operation.

Operational on both Classified and Unclassified Levels

The ASSWAP program functioned on both classified and unclassified levels. It was operated jointly out of the Pentagon and at least for a time at the underground complex in Las Vegas owned by Bigelow Aerospace.

Recent disclosures by the Defense Intelligence Agency, have confirmed that they were testing wreckage recovered from UFO crashes at Bigelow Aerospace in the underground complex which was obviously highly classified.

AATIP, through a contract awarded to BAASS (Bigelow Aerospace Advanced Space Studies) generated a 494-page **classified report** that documents alleged worldwide UFO sightings over several decades. This Report has not been released to the public but focuses on UFO reports, plans and extensive analysis of unexplained aerial phenomena.

The BAASS report was only a small part of the materials provided to the Defense Intelligence Agency as monthly reports were being sent to the Pentagon in addition to annual program updates that were all about UAP or anomalous phenomena.

The program also funded and published 38 studies some of which have recently been declassified. Those studies cover a range of advanced, exotic and theoretical aerospace topics ranging from:

- High-Resolution Tracking at Hypersonic Velocities
- Quantum Entanglement Communication
- Warp Drives
- Wormholes and Space Time
- Invisibility Cloaking
- Dark Energy
- Manipulation of Dimensions
- Negative Mass Propulsion

BAASS / MUFON SIP Project (Unclassified)

In February 2009 an agreement was made between BAASS and MUFON whereby MUFON provided data from the sighting reports in exchange for funds paid each month from BAASS. The project was referred to as the **BAASS / MUFON SIP Project.** (SIP = Star Team Impact Project)

As indicated, I served as the Chief Investigator for MUFON in both Virginia and North Carolina from 2005 to 2009. I also managed the Star Team, which was a small select group of highly experienced investigators who worked on the most compelling cases.

In 2009, I was hired to serve as the Manager of the **BAASS / MUFON SIP Project.** This was the most advanced rapid response UAP/UFO Civilian Investigative Team in the world; although it was funded through a secret contract with ATTIP it was **not** classified. I supervised this Project from February 2009 until February 2010 when it ended.

The BAASS / MUFON SIP Project was in part conceived to provide funding to professionally investigate UFO/UAP events. **Investigators** were compensated for their time on deployments and all expenses were paid by the project.

Part of this funding was also allocated to payroll a team of full time **Dispatch Operators** who would work continuous shifts throughout the week and monitor the incoming sighting reports on MUFON's Case Management System. There were approximately 800 UFO sighting reports being submitted to MUFON each month and about 35 to 50 of them would be selected by the dispatch operators as significant cases.

I reviewed significant cases each month and assigned investigators to deploy to the locations, evaluate information, interview witnesses, and write reports. Then I edited the Final Investigation Case Reports and forwarded them to BAASS. A lot of the information in the monthly reports that BAASS was sending to the Pentagon came from data I prepared and submitted to BAASS each week.

During that year, while managing the BAASS / MUFON SIP Project I learned an incredible amount about how to investigate and study this Phenomena. I developed long-term relationships with some of the most experienced UFO Investigators, Researchers and Ufologists in the world.

The "Catbird Seat" Keep in mind we were receiving 650-850 case reports per month and we were literally looking at this information night and day (7 days a week) for about a year.

Metaphorically speaking this was the "Catbird Seat" if there ever was such a thing in the Ufology Community. Throughout that year, I took advantage of every minute, to study this phenomenon and learn all I could about how to investigate sighting reports and how that data might be effectively used.

Documentation of my work

In October 2021 George Knapp published a book "***Skinwalkers at the Pentagon***" with James T Lacatski and Colm A Kelleher who I worked with during the BAASS SIP Project. This was marketed as an insiders account of the secret government UFO Program. During the time I worked on the BAASS / MUFON SIP Project, Bigelow owned the Skinwalker Ranch. This book is excellent and references some of the investigative work that was done on the ranch and the paranormal aftereffects, as well as a lot of information about the BAASS / MUFON SIP Project. I am very grateful to the authors of this book because they referenced me by name and acknowledged the work that I did on the project.

A second book ***"Inside the US Government Covert UFO Program: Initial Revelations"*** by George Knapp, James Lacatski, and Colm Kelleher was published in 2023 about

AAWSAP, BAASS and MUFON. In this book they referenced me in the following excerpt:

> ***"Richard Lang was hired by MUFON as the STAR Team Impact Project (SIP) Coordination Manager with an effective hire date of March 9, 2009. STAR stands for Strike Team Area Research, a specialized paid team of investigators who could be rapidly deployed to the most compelling UAP cases . . . The excellent performance of Mr. Lang and his STAR investigators contributed greatly to the success of AAWSAP operations"***

STATEMENT TO CONGRESS

Submitted by George Knapp to the House Oversight Committee's Subcommittee on National Security.

George Knapp is the chief investigative reporter for KLAS TV in Las Vegas. KLAS is Nevada's original television station and a CBS affiliate. He is a journalist who has had an interest in UFO secrecy beginning in 1987. In the years since then he has written hundreds of UFO related news stories and series, probably more stories over a longer period of time than any other mainstream journalist in the country. I have known George since 2008 through my association with BAASS and have been interviewed by him three times on Coast-to-Coast AM Radio and on his podcast *Mystery Wire*.

In July of 2023 George submitted a formal written statement to the House Oversight Committee which was accepted in the Congressional Record. This complete statement is included in the appendix of this book. He outlined his years of work investigating UFO phenomena and the secrecy that surrounds it. There's significant detail about Bigelow Aerospace and the research they've done over the last 20 years.

My Work as an Author and Publisher

This is the third book I have published on the subject of UFO Investigation. In my first book ***"UFO Investigation the Methodology for a New Age"***, chapter four is the largest chapter in the book and includes a very detailed discussion about the BAASS / MUFON SIP Project and STAR Team.

The book also includes Methodology (methods, techniques, procedures) that are used to facilitate UFO/UAP Investigation.

The book was written to provide advance training for UFO investigators and has several chapters on the use of video in investigation and examples of equipment that investigators need to have and learn how to use.

There is considerable discussion about other associated anomalous phenomena to include: the anti-physical effects (observed behavior inconsistent with current physics), close proximity encounters with individuals, physiological effects on humans, psychic effects in the realm of human consciousness as well as other high strangeness which is explored in that book.

One final note about the information in that book; I had made significant effort to carefully document my training, experience, associations, and certifications that helped me to facilitate this research. I have found my share of breadcrumbs along the way, where individuals knowing about certain technology who could not directly talk about it gave me hints on where to look for information to study which helped me understand certain types of phenomena.

With that said there are some concepts in that book (and in this book) that I have described and noted, particularly in the technical fields involving high strangeness that I have specific knowledge of and clearly understand. However, I will not explain how I know and understand these concepts or disclose the source of the information that I was able to find.

Many of my certifications mentioned have expired and are no longer legally valid. However, they serve to support my education, experience, and training in my evolution as an authority in investigation methodology.

END CHAPTER

Chapter 6
Problems with Disclosure and Declassification

Let's do an analysis of the problems that will be associated with Disclosure and Declassification.

1) False Flag

A False Flag Operation is an act committed with the intent to disguise the actual source of responsibility and pinning blame on another party. The term today extends to include countries that organize covert attacks on themselves and make the attacks appear to be committed by enemy nations or terrorists thus giving the nation that was supposedly attacked a pretext for domestic repression and foreign military aggression. [13]

Many in the UFO community including a surprising number of government insiders seriously believe that the current efforts to declassify and disclose information about UFO's is essentially a precursor for an orchestrated well planned False Flag Event.

Funding Planetary Defense

The thinking is that the Defense Contractors in the Military Industrial Complex will covertly create what appears to be an attack by some Extraterrestrial civilization. There are various scenarios but the most common is that some type of a holographic projection might be displayed in the sky to simulate the presence of huge Extraterrestrial ships that threaten to attack cities around the Earth.

Regardless of how this threat is conveyed, the tax coffers will obviously be tapped and opened by congress and the Defense Contractors will soon be allocated billions of dollars to invest in new weapons to defend the planet.

Must Acknowledge They Exist

The point being that before they can start pointing the finger at the Extraterrestrial enemy (bad guys), they must first come forward and admit they actually exist which is what may be happening with declassification.

The Pentagon has been starting to release information about UFOs through an organization known as TTSA that we will take a look at in the next chapter. TTSA has a membership of career officers from the NSA (National Security Agency), the CIA (Central Intelligence Agency) and DIA (Defense Intelligence Agency). It is interesting that these same agencies who for the last 70 years have been involved in the UFO coverup and keeping the secrets are now possibly involved in the disclosures.

With the advent of fake news and the power that the broadcasting companies and social media have over the general population, a false Flag operation could actually be a possibility.

The Panic Hazard

The disastrous effect that UFO activity could have on the general population is the creation of fear, panic, threat, and all kinds of irrational behavior. There is a well-documented case of mass hysteria created by **War of the Worlds,** the radio drama by Orson Welles about an invasion of Martians.

Orson Welles 1938 Radio Broadcast

It was a CBS radio broadcast on Halloween of 1938. "The drama was realistically presented as several news bulletins describing an alien spaceship landing in New Jersey. Then alarming news updates detailing a devastating alien invasion taking place around the country and the futile efforts of the U.S. military to stop it, while the fictional radio news reporter was pretending to be choking on poison gas as the Martians overwhelmed New York.

NY Police actually entered the CBS studio and a struggle evolved with thc CBS executives who were trying to prevent the cops from busting in and stopping the show.

The radio broadcast resulted in many hysterical actions, including thousands of panic-stricken phone calls to police stations, scattered reports of listeners fleeing their homes, massive traffic jams, and wildly driven automobiles trying to escape the cities. The result was a massive panic scenario and a nightmare for the police and public officials.

Power Failures and UFOs

Numerous power failures have been reported in association with UFOs in Brazil (1957), Rome (1958) and Mexico (1965).

"The most notorious of all blackouts associated with UFOs was the Power blackouts that fell over 30 million people in the northeastern corner of the U.S. during the early evening rush hour period on November 9, 1965.

Power relay services that were supposed to automatically transfer power in case of a failure from one area to an alternate source, all simultaneously malfunctioned. Military communications relying on public power without alternate backup systems also failed but media communications were operable to make a quick public announcement that there was no military emergency.

Though it was largely over by the next morning, the official explanation about a malfunctioning device in a Canadian hydroelectric generating plant never accounted for the failure of millions of dollars' worth of electronic devices to shift the load when the breakdown occurred.

Military Installations

Many researchers believe that UFOs are here on a surveillance mission; the fact that a majority of sightings occur around our military installations and research and development areas leads to the conclusion that a methodical study is being made of the Earth and its defensive and offensive capabilities. There is the testimony of numerous military personnel who have indicated that huge UFOs have hovered over missile silo complexes and deactivated nuclear weapons on more than one occasion. UFO researcher Robert Hastings who organized the National Press Club briefing said more than 120 former service members had told him they had seen unidentified flying objects near nuclear weapon storage and testing grounds.

Space Exploration

Because of our recent adventures into space there are some who speculate that UFOs are more concerned with what we will do out in space. In any event, the Air Force's official publication (issued by the Government Printing Office 1968) titled Flying Objects says that "No UFO has been determined to represent a threat to our National Security. That claim is rapidly changing as we will see in the next chapters, as now UFOs are considered a threat to National Security.

They Would Already Have Obliterated Us by Now

Looking at the evidence of extraterrestrial contact and their impact on civilizations and societies around the planet, It's going to be a pretty hard sell to say that they have malicious intentions. In my thinking if they were here to harm us they would have already done it a long time ago, trust me they have technology and weapon systems that are 100s of years ahead of us. Keep in mind that there are a number of different species that have been coming here for a long time and they have different agendas. But in my research most of the communication that has been established with the abductees (CE5) that I have interviewed would indicate that Extraterrestrials (ETs) intentions are for the most part benevolent.

2) Adverse Consequences of Declassification

A great example about declassification and the resulting adverse consequences reminds me of when the secret underground bunker underneath the **Greenbrier Hotel in West Virginia** was declassified. This was a state-of-the-art underground facility built to house 300 members of congress in the event of a nuclear strike on Washington, DC.

The bunker was constructed under top secret orders during the Eisenhower administration (completed construction in 1962) with funding from secret government black project funds. It was done during a building expansion project when they added a new wing on the hotel above ground (bunker was secretly being built underneath the new wing). It operated for about 30 years in a state of stand-by readiness to accommodate the Congress if necessary.

The Bunker was a massive expansion to the hotel building (addition of over 112,544 square feet underground) and a secret staff manned the bunker 24 hours a day. The money to build all this and pay the employees' working underground was laundered through the Railroad company that serviced the town (to keep everything secret).

The Washington Post ran an article in 1992 disclosing the existence of the bunker (blowing their cover) and ultimately the facility was declassified by the US Government.

Since the Bunker and the employees working there did not exist (officially) most of the tax paying obligations were obviously ignored. However, after the declassification, the issues about taxes that had not been paid (including 30 years' worth of property tax on prime real-estate) immediately came up and was investigated by the state and county governments which ended up with lawsuits, tax liens and plenty of dirty laundry to clean up there.

I am sure that the officers of the Military Industrial Corporations involved already know this is going to be a nightmare for them when and if they must come clean with all they have been involved in. They will likely hang on to the private proprietary information and patent rights they own for as long as they can.

3) Adverse effects on the stock market

What if we met for lunch and I brought with me a large briefcase or a small suitcase and set it on top of the table where we were eating. Then I had advised you that the device inside this suitcase could generate electric to power every house within a 100-mile radius of this table, that could operate for 100 years without any renewable energy and it would not generate any type of pollution or nuclear contamination, not to mention that it's a free energy source that really can't be metered or sold.

In 1998 I met for lunch in Charlottesville, with a distinguished Doctor who posed that same question to me. At that time, I was a Senior Officer of the Trust and Investment Management Division of a large commercial bank and a significant part of my work involved stock market transactions and Investment Management for clients.

My answer was somewhat spontaneous and I said, "well our economy is predominated by the petroleum industry and such a device would literally wreck the stock market and economy" He agreed and we enjoyed our lunch in a little wine shop and restaurant on the downtown mall talking about Zero-Point energy.

Bottom line this is just a small example of the technology that's been developed over the last 70 years through back engineering projects. Extremely efficient energy sources like this have been developed through engineering processes, by studying the energy systems in these recovered crafts wreckages. Sadly, government agencies and energy corporations, predominantly from the petroleum industry, have carefully procured the patent rights on these types of devices and secured them from public view. They are not going to let the public have these types of energy sources until they sell the last gallon of gasoline and oil.

We could go on here for another 100 pages talking about the advanced medical technology that has been developed and suppressed by the pharmaceutical industry or Anti-gravity propulsion systems that are only available to the most advanced secret defense contractor operations. We could go on and on, but I'm sure you get the point.

4) Lost in Bureaucratic Red Tape

The Senate Intelligence Committee's 2020 demand for a public report about the government's UFO knowledge, was officially signed into law on December 27th, 2020, and the

180-day deadline arrived in late June, just in time with the publishing of this book.

America's 3-letter agencies (NSA, DIA, CIA) to name a few, do not share information and do not like to work with each other. In the Intelligence Industry, information is powerful, and data is deliberately segregated and cordoned off from rival agencies. This approach compartmentalizes the U.S. Security Agencies and leaves an inadequate picture about what is behind reports particularly related to UFO/UAPs.

This is a big problem as we will see when we analyze the Assessment Report. The compartmentalization of information that occurs when different security agencies try to hide information from each other explains in some respect the lack of information conveyed in the Assessment Report

The civilian UFO community seems to be expressing concerns that Disclosure Information will ever see the light of day. Many observers also predict that Pentagon redactions will conceal any significant revelations. The NY Times has already started claiming the report doesn't reach any definitive conclusions about the origin of the objects. The former Director of National Intelligence John Ratcliffe indicated that the report will contain details about "things that we are observing that are difficult to explain."

While skepticism about the congressional UFO/UAP Assessment Report is justified, there are plenty of reasons to look forward to its public release. The committee mandated a detailed analysis of Unidentified Aerial Phenomena data and intelligence reporting which means the agencies are required to provide explicit details and interpretations about the UAP situation. The Assessment Report contains new revelations for UFO investigators to explore which we will analyze in Chapter 8.

5) Advanced Technology Held Privately

Remember, most of this research and the technology that has been developed over the last 70 years was facilitated through private multi-national corporations and defense contractors as opposed to working directly in government agencies and on military bases, which obviously makes their research very private and exempt from FOIA requests.

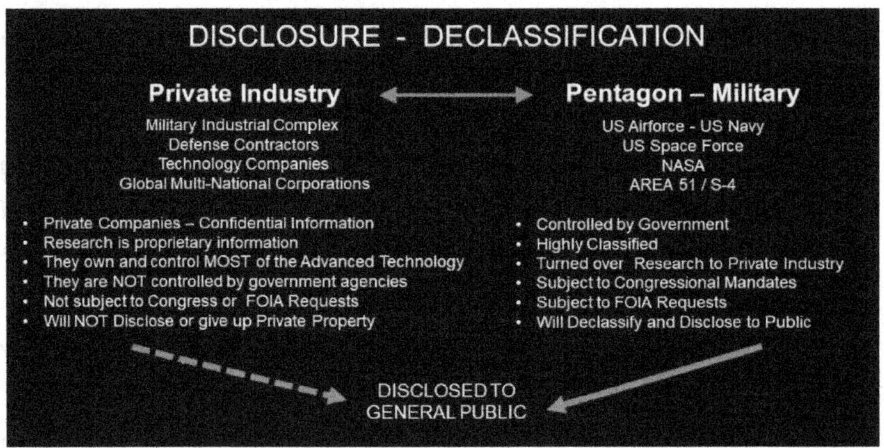

According to information obtained in late 2020 through a FOIA request from the Defense Intelligence Agency, the Pentagon has publicly admitted to testing wreckage recovered from UFO crashes indicating some of this material was placed with a defense contractor for analysis and storage in "specialized facilities". The documents revealed the tests were carried out by Bigelow Aerospace, in the underground facility at Las Vegas and that the company performs private contract work for the Department of Defense.

The key words here are: **"private contract work for the Department of Defense"**. During my time working on the BAASS / MUFON SIP Project, BAASS did not share findings or test results with us, as the deal clearly stated that BAASS was a private organization, and their research was proprietary information. I am sure that most of the other defense contractors and technology research corporations operate the same way and have similar agreements in place.

On the other hand the Military Industrial Complex, including the defense contractors and research corporations are going to be disinclined (not in any hurry) to release their private proprietary information.

Think about all the money they have made using funds that were funneled to them through black budgets and secret government spending programs. The public is going to want accountability for the public funds and taxes, as well as disclosure of advanced technology that was paid for with US tax dollars and hidden from them. This will undoubtedly open the door to massive lawsuits and in some cases with legal action against the officers of some of the corporations.

6) High Strangeness Associated with UFOs.

When the movie ET was produced by Steven Spielberg an event was held by President Ronald Reagan at the White House to premiere the new movie. During the premiere of the movie at one point President Reagan made the remark to several people close to him saying “there are only six people in this room who know how true this really is”.

Public Perception of Reality

During the early stages of our lives as a result of our education, family situation and our environment we all develop a perception of reality, what we believe is possible (or not possible) and how the basic laws of physics affect our lives and the way we live.

Studies indicate that by age seven we actually learn almost half of everything we will learn in our life. Think about it, shortly after we are born, our vision and sensory perceptions develop. We learn to see, feel, taste, smell, touch and feel pain. By age two, we begin to talk and walk. Then we begin to interpret facial expressions, learn how to express emotions and ultimately communicate with those in our environment.

Next of course we learn the rules. The rules are typically made by the people who raise us, care for us, and control us. By age seven we are fairly well locked in by the ideological concepts of our parents, our schools, religious organizations, government agencies and peer pressure of those around us.

Most important as we get a little older, we begin to understand the constraints of our three-dimensional physical reality. By constraints I mean physical space (length-width-depth) and most importantly time. Psychic research centers and most religious organizations will ask the perspectives to acknowledge that they are more than their physical body. The concept of consciousness may be how we define and understand our own existence within our three-dimensional physical reality and the non-physical part of our existence.

Regardless of where you were born or how you were raised, you have been programmed. You have conscious presence of mind and understand who you are, where you live, what you do, how much you are worth and who your friends, family and enemies are.

But on a subconscious level you have been subject to immensely powerful programming, in terms of your belief systems and what you think is possible (or not possible).

Very High Strangeness

Part of the difficulty one encounters when trying to understand the UFO phenomena is that there are many instances of what we call very high strangeness, which dramatically adds to the difficulty in investigating these cases and writing objective rational reports about what occurred.

Looking at the anti-physical characteristics and some of the psychic effects that occur during UFO sightings and close encounters will give the reader an idea as to the complex nature of this phenomena, and the difficulties that will be encountered when trying to disclose it to the general public.

Anti-physical Characteristics of UFO encounters

Anti-physical for lack of something better to call this, includes objects (UFOs) that may be physical, but they completely defy the current laws of physics in events of very high strangeness. Some examples are listed herein. [14]

- A Transmedium Vehicle is an object that moves through multiple mediums, such as air, water or space with no displacement or a sonic boom. Also they have been tracked on sonar traveling at incredible speed underwater.
- Transmedium vehicles sometimes appear to be plummeting into the solid ground very smoothly without impact or crash, like a rock that is slipped into water (also appear to travel into the side of a mountain without the presence of any opening or doorway).
- Orbs of light observed appearing as balls of colored, intensely bright light sometimes pulsating and under intelligent control.
- Crafts disappearing at one point and appearing elsewhere instantaneously.

- Affecting observers who experience missing time (time dilatation); for example a witness on their way home from the mall has an encounter with a craft at close proximity that appears to only last for several minutes. Upon arriving at home, which is 10 minutes from the mall they realized that 3 hours have passed since they encountered the object.
- UFOs near automobiles can cause engines to stop and electrical systems to completely fail. Witnesses indicate that during such an event the engine stops running, all the electric power goes out (headlights, interior dash lights, radio etc.) and cellular phones go dark. After the event suddenly the electric power and lights are back on and the engine is running.
- Levitation of the witness, vehicles and animals in the vicinity of an encounter.
- Producing topological inversion or space dilatation (object was estimated to be of small exterior size, but when thc observer entered they experienced a huge interior many times the exterior size).

Inter-Dimensional Portals

Another element that should be included with the Anti-physical layer that remains worth mentioning is Inter-Dimensional Portals. Inter-Dimensional Portals are openings which are presumed to be passageways into another dimensional reality which are often associated with missing persons disappearing in the wilderness and national parks.

- Starts with a machine like feeling or noise, sometimes described as a whirling noise which is the opening of the portal.
- Victims see something blurry (distorted image) in front of them on the trail that is interesting and attractive but not scary to them.
- Sometimes these encounters include an entity that is tall, slender and appears to be female, who lures people into the portal.
- Induction through the portal opening is swift and powerful but silent, and sometimes causes shoes and other articles of clothing to be ripped from the individual being pulled in.
- After the event when the portal closes violent weather including thunder and lightning storms occur.

Individuals enjoying the national forests and hiking sadly do disappear without a trace in alarming numbers. Typically, these individuals are alone when they disappear, having recently walked away from the other members of their hiking group.

Searchers find shoes, boots and articles of clothing near where they were last seen but no trace of a body. Sometimes after several weeks the victim's body is found 10-12 miles away without rational explanation as to how they got there, other times they are never seen again,

Inter-Dimensional Portals are typically connected to a geographic area whereas Abductions are usually associated with some type of UFO craft that is hovering or has landed nearby.

Abduction

Abduction is similar in the sense that individuals are pulled into higher frequency dimensional existence with lower density but usually confined inside a craft and later returned to their original location. Many experiencers (people that have experienced abduction phenomena) talk about being in their homes or near their house and some even in their bedrooms and they remember being taken away by entities, to be examined on some type of craft and then eventually returned unharmed with the exception that the psychic affects may bother them for the rest of their life.

- They hear a low frequency, humming sound.
- They feel a vibration in their body.
- They experience a lack of mobility (paralyzed).
- They are confronted by some type of entities that take them away, sometimes right through the walls or ceilings, into a craft which is usually hovering in the vicinity of where the abduction takes place.
- They indicate the craft is much bigger on the inside than it appeared to be on the outside.
- The buzzing sound is associated with being surrounded by higher frequency energy and actually being pulled into a higher dimension which would explain the reduction of density allowing them to pass through solid objects.
- Experiencers typically will indicate that they were put through some type of a physical examination while on the craft.

- Sometimes they were in the presence of other humans that were taken as well.
- Loss of volition experienced during and after events.

The underlying rationale for being taken varies; some experiencers, particularly females indicate that they have undergone some type of a gynecological exam and in some cases after the event when returning home they experience the symptoms of a miscarriage.

Psychic

The psychic category of effects can only be labeled psychic because it involves a class of phenomena commonly found in the literature of parapsychology. Sighting reports contain contain accounts of these effects. [16]

- Impressions of communication without a direct sensory channel sometimes known as mental telepathy.
- Poltergeist phenomena: motions and sounds without a specific cause, outside the observed presence of a UFO.
- Maneuvers of a UFO appearing to anticipate the witness' thoughts. This is commonly reported where essentially a witness will indicate that they've gone out to smoke or let the dog out late at night and they notice a light at great distance and high altitude. After observing the light for a period of time the light seems to anticipate their attention and zooms in on them ultimately being very close to directly over them.

- Premonitory dreams or visions (a dream that appears to give advance notice or warning of some future event).
- Personality changes promoting unusual abilities in the witnesses like clairvoyant abilities.
- Healing - witnesses report recovering from illnesses after encounters and also some indicating that they can heal other people.

Looking at a report where a witness claims that the object could anticipate their thoughts; on the surface this sounds unbelievable, however when these same phrases are articulated in hundreds of sighting reports, it makes an interesting case that such phenomena may have actually occurred.

Multi- Dimensional Existence

Ancient Teachings and recent Quantum Physics both indicate and define the existence of 12 dimensions. There is much written about this research in the world of Quantum Physics and the search for other dimensions.

Dimensions are different planes of existence defined by a Vibratory Rate. Each dimension has defined parameters of existence and constraints that are in tune with the frequency of that dimension.

It is like when you turn on a television and select a channel, set at a certain frequency. You can watch a program and then change the channel by changing it to a different frequency. Both channels come in clearly when the frequency is set by the channel selector. Both channels exist simultaneously but you can only experience the one which is set to the frequency you are tuned to and watching. The Dimensions all exist simultaneously but the observer can only experience the plane of existence that is in tune with their frequency.

Many experienced UFO /UAP researchers believe that these objects and vehicles can somehow change their frequency and move in and out of our 3-dimensional time space reality.

Turning the Corner

I would be the first to admit that during the early days of my exploration I saw things and worked on cases that I did not understand completely. My scientific education and training in Aeronautics and my experience as a commercial pilot were obviously formative in my thinking.

My perception of reality has drastically changed since that flight in 1976 and I look at case report data much differently now, partly because I have had the chance to analyze hundreds of witness case reports and interview many people associated with abductions and close encounters over the years in the course of my research.

Consciousness

In University physics class back in the 1970's we were taught that Space, at least from an interplanetary perspective was a vacuum. Travel outside of the atmosphere of the earth space was simply empty of any substance and void.

What we know now is that the universe is connected by a fabric or energy that occupies all space, which is consciousness; space is not void as originally thought. Whether you study ancient teachers or quantum physics the understanding is the surprising similar.

Consciousness is the missing element in exploring the UFO/UAP phenomena. In our effort to understand the phenomena we study the types of vehicles, try to understand the inter-dimensional method of travel and the energy sources used to power these incredible crafts. Trying specifically to understand who these entities are, where they come from and why they come here.

7) Final thoughts about Problems with Disclosure

Throughout my 30 years of UFO research the information contained in this chapter and the occurrences that are described herein are real and actually do happen. I've looked at literally hundreds of UFO reports over the years and as you can see, some of this research involves extremely high strangeness that's difficult to convey to the average person.

Laying this all out in a blatant disclosure program, with no holds barred for the general public, is not going to be a practical way to facilitate this transition to full disclosure.

More about this in Chapter 9 see War of the Worlds Season Two Coming to your TV News Network soon!

The element of fear is the most controlling factor, a program like this would be devastating, causing panic and chaos like it did in 1938 as well as adversely affecting the public confidence in government.

Steps In Systematic Release of Information

This is why I believe that a systematic release of information will need to be in the works through a series of steps that will ultimately lead to full disclosure.

In the final chapter of this book, we're going to take a look at those steps and how that information might be carefully released to the general public over time in a way that dispels anxiety through a better understanding of quantum physics and human consciousness.

Hopefully, this chapter will bring the readers to a better understanding of the issues and problems that will be associated with Disclosure and Declassification.

END CHAPTER

Chapter 7
News Releases and Events
Leading to the Preliminary Assessment Report
Chronology of events that are featured in this chapter:

- In 2008 the Pentagon established the Advanced Aerospace Weapon Systems Applications Program (AAWSAP) which was operated and funded by the Defense Intelligence Agency. This program funded the BAASS MUFON SIP Project (discussed in Chapter 1).
- United Nations officially appoints an Ambassador to Extraterrestrials (for the Alien World) in the event they contact humanity, November 2011.
- TTSA (To the Stars Academy of Arts & Sciences) was established in 2017.
- CBS 60 Minutes interviewed Robert Bigelow, May 2017 .
- Washington Post - December 2017, The Pentagon's secret UFO Program was exposed on December 16.
- In 2017 Navy Fighter Jet camera footage of UFOs was leaked to *Politico* and *New York Times* which drew massive public attention to the UFO subject.
- 2019, The Pentagon confirmed the authenticity of newly leaked video and images showing multiple UFO sightings by U.S. Navy personnel.
- By 2019 members of Congress were receiving classified briefings about UFOs.

- In 2020 the Senate Intelligence Committee led by Marco Rubio mandates the Pentagon provide a report about UFOs.
- Later in 2020 the Pentagon established the UAP Task Force (UAPTF).
- The Appropriations Bill signed on December 27, 2020 mandated that a report detailing what the government knows about UFOs be released this year.
- The UAP Task Force in coordination with the Office of the Director of National Intelligence (DNI) are leading the investigation and compiling the report for Congress, a report detailing everything the government knows about Unidentified Flying Objects.
- June 18, 2021, Members of the House Intelligence Committee received a Classified Briefing prior to the release of the UAP Assessment Report.
- July 25, 2021, **UAP Assessment Report** was published. (see Chapter 14)**.**
- **National Defense Authorization Act (NDAA) of 2023.** People with information about UFO / UAPs have been given amnesty from prosecution. (see Chapter 14).
- **January 2024 - House bill would allow civilian pilots to report UAP sightings to the FAA.** Rep. Robert Garcia (D-California) and Rep. Glenn Grothman (R-Wisconsin) sponsored a new bill, which would create a reporting mechanism for civilian pilots who claim to see UFOs. Under the provisions of the bipartisan legislation that Garcia and Grothman

proposed, civilian and commercial pilots would be able to report UAP sightings to the Federal Aviation Administration which would be compelled to raise them up to the Pentagon's All-domain Anomaly Resolution Office. FAA air traffic controllers, flight attendants, maintenance workers, dispatchers, and airlines themselves would also be able to make UAP reports under the legislation. The Bill includes legal protections for those who report sightings. In the proposed legislation is language providing legal protection to those who make reports.

Featured Articles and Excerpts

Unidentified Aerial Phenomena Task Force (UAPTF)

The Pentagon's Unidentified Aerial Phenomena Task Force is now investigating these unidentified aircrafts. The UAPTF was formed in the summer of 2020 in an effort to improve the Department of Defense's "understanding of, and gain insight into the nature and origins of UAPs" particularly those "incursions by unauthorized aircraft into our training ranges or designated airspace.

As DOD has stated previously, the safety of personnel and the security of operations are of paramount concern. The Department of Defense and the military departments take any incursions by unauthorized aircraft into our training ranges or designated airspace very seriously and examine each report. This includes examinations of incursions that are initially reported as UAP when the observer cannot immediately identify what he or she is observing.

Intelligence Authorization Act for Fiscal Year 2021

As indicated, part of the $2.3 trillion dollar Appropriations Bill signed on December 27, 2020, mandated that a report detailing what the government knows about UFOs be released this year. The UAP Task Force in coordination with the Office of the Director of National Intelligence (DNI) are leading the investigation and compiling the report for Congress.

So, what is going on?

Over the last 4 years it seems that the Pentagon is starting to release information about the Government's involvement with UFO/UAP activity which many believe was staged in part by TTSA.

TTSA (To the Stars Academy of Arts & Sciences).

TTSA was established in 2017 as a public benefit corporation to promote topics such as UFOs and studies about Consciousness, Telepathy and Psychokinesis and supports research into explaining the technical advances these reported UFOs demonstrate. Their goals to generate funding to underwrite significant research in these areas and develop collaborative relationships between Government, the Aerospace Industry and Academia.

None of the members of TTSA consider themselves Ufologists' or part of the Ufology culture. They claim that most of them come from a U.S. Government background, both Defense and Intelligence.

TTSA was initially founded by:

- **Jim Semivan**, a former senior Intelligence Officer with a 25-year career in the CIA.

- **Harold E. Puthoff**, a Senior Adviser to the AATIP program, his background is with NSA, and he was director of CIA/DIA funded research programs.

- **Tom DeLonge,** an entrepreneur and award-winning American musician.

- **Lue Elizondo** who ran the ATTIP program at the Pentagon resigned to join the group in 2017. According to Lue Elizondo, the mission of AATIP was quite simple. It was to collect and analyze information involving anomalous aerial vehicles. When Elizondo took over in 2010, he focused on the National Security implications of unidentified aerial phenomena documented by U.S. service members.

- **Christopher Mellon** 20 years U.S. Intelligence Community as the Minority Staff Director of the Senate Intelligence Committee and the Deputy Assistant Secretary of Defense for Intelligence. He conceived and drafted legislation establishing the US Special Operations Command in 1986.

Other members included Senior Intelligence Officers from the Department of Defense and the CIA, including Steve Justice formerly director of advanced systems at Lockheed Martin Skunkworks to name just a few.

TTSA Negotiations with Media Outlets

Members of TTSA claim to be responsible for legitimizing the UFO topic in the mainstream press through their special relationships with the New York Times, Washington Post, CNN and Fox News.

Specifically, they are claiming responsibility for the information leaked to the press and the articles on the next few pages from New York Times and the Washington Post. Also taking responsibility for the release of the high quality F18 fighter jet videotapes of UAP encounters from the Pentagon Archives that had been made available to the public, while at the same time putting their spin on all this indicating that UFO's may be a threat.

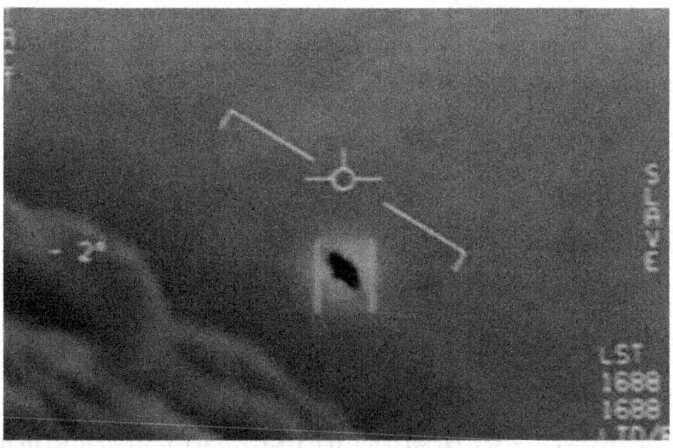

TTSA Financial Issues

In September 2017, the company (TTSA) began offering $50 million worth of public stock through a Regulation A+ Equity Crowdfunding Campaign. According to SEC filings, as of October 2018 only $1 million of those shares had been sold and the company had a $37.4 million deficit largely from a stock incentive plan funded for its employees which has brought into question the TTSA's financial sustainability.

May 2017, 60 Minutes interviewed Robert Bigelow.

CBS 60 Minutes broadcast an interview with Robert Bigelow in May of 2017 which was titled *"Bigelow Aerospace founder says commercial world will lead in space".* [8]

During the CBS 60 Minutes interview when Mr. Bigelow was asked by correspondent Lara Logan, "Do you also believe that UFOs have come to Earth?" he responded, "There has been and is an existing presence, an ET presence" . . . and "I spent millions and millions, I probably spent more as an individual than anybody else in the United States has ever spent on this subject."

Washington Post - December 2017

The Pentagon's secret UFO program was exposed on December 16, 2017, by the Washington Post in an article titled: *"Head of Pentagon's secret 'UFO' office sought to make evidence public"*, written by Joby Warrick. [16]

This article was about an intelligence officer who arranged to secure the release of unusual videos in the Pentagon's secret vaults including raw footage from encounters between fighter jets and Anomalous Aerial Vehicles (AAV) all taken from fighter jet cockpit cameras showing pilots struggling to lock their radars on oval-shaped vessels.

According to the article, an internal Pentagon memo requesting that the videos be cleared for public viewing, it was argued that the images could help educate pilots and improve aviation safety, subtly conveying a threat to aviation.

New York Times - December 2017
Then an article published in the New York Times by Helen Cooper, Ralph Blumenthal and Leslie Kean, *"Glowing Auras and 'Black Money': The Pentagon's Mysterious U.F.O. Program.* Bottom line is the existence of the program, known as the Advanced Aviation Threat Identification Program (AATIP) was confirmed officially for the first time by the Pentagon in these articles. There was Video embedded (included links) in the articles that show fighter jet cockpit cameras recordings. [16]

November 2004 USS Nimitz Encounter Disclosed.
Some of the Videos included in these articles show encounters between Navy F-18 Super Hornets and AAVs.

The USS Nimitz Carrier Strike Group, which included the nuclear-powered aircraft carrier and the missile cruiser USS Princeton, were conducting a series of drills while located in the Pacific Ocean about 100 miles southwest of San Diego, California.

As indicated, AAV is a term the military prefers to use as opposed to UFO or UAP. The Princeton's radar team had been identifying numerous AAVs appearing at altitudes greater than 80,000 feet, far higher than commercial or military jets typically fly, and they began detecting instances in which the AAVs dropped with astounding speed to lower, busier commercial airspace.

On November 14, 2004, Two F/A-18F Super Hornet Fighter Jets from the Nimitz that were already airborne doing training received an order from an operations officer (air controller) aboard the USS Princeton. They were told to stop their training maneuvers and proceed to new coordinates to investigate an AAV.

The Fighter Pilot of the USS Nimitz Carrier Strike Group and his wing man (who was actually a female fighter pilot) were vectored to the location of possible AAVs. While enroute they were asked if they were carrying live weapons and they replied that they were not.

The Fighter Pilot reported that he saw an unknown object, white and oval like a Tic-Tac, hovering above an ocean disturbance. He described it as about 40 feet long, shaped like a Tic-Tac candy and with no obvious means of propulsion: "It's white. It has no wings. It has no rotors".

As he flew around it, he says the craft ascended and came right at his plane: “All of a sudden it kind of turns and rapidly accelerates, beyond anything I’ve seen, crosses my nose, and…it’s gone.

Several videos of these Tic-Tac’s were filmed with cockpit cameras in the fighter jets. These F18 cockpit videos were filmed with Infrared FLIR (Forward Looking Infrared) cameras which would show any type of exhaust gas discharge from propulsion on the video. This was the not the case as these objects seem to have no evidence of propulsion, exhaust gas, or any aerodynamic characteristics like wings or rotors.

March 2019 - UAPs flying over Oceania

The "ACORN": An FA-18 Pilot and a Weapon Systems Officer took these photos of Unidentified Aerial Phenomena Courtesy: George Knapp/Mysterywire.com.

July 2019 - UFO off the San Diego Coast

An 18-second video shows what is described as three pyramid-shaped UFOs hovering over the warship USS Russell at night. At one point the pyramid-shaped crafts reportedly hovered 700 feet over the tail of the Russell.

Pentagon Spokeswoman Susan Gough

The Pentagon has confirmed the authenticity of newly leaked video and images showing multiple UFO sightings by U.S. Navy personnel.

Pentagon spokeswoman Susan Gough stated "I can confirm that the referenced photos and videos were taken by Navy personnel. The UAPTF has included these incidents in their ongoing examinations. "As we have said before, to maintain operations security and to avoid disclosing information that may be useful to potential adversaries, DOD does not discuss publicly the details of either the observations or the examinations of reported incursions into our training ranges or designated airspace, including those incursions initially designated as UAP."

Note: These videos were released by the Defense Department's Advanced Aerospace Weapon Systems Applications Program (AAWSAP) in cooperation with the TTSA officers as indicated. The Pentagon's candor and decision to even acknowledge that the video and images of pyramid shaped aircrafts and UFOs over waters near Oceana are in fact real and is energizing UFO researchers and experts worldwide who hope it is the beginning of more transparency.

This information is authentic, but it is only the tip of the iceburg. The Public Disclosure from these two articles has helped in bringing the UFO subject back to the front burner in the public view. Links to the articles can be found in the reference section at the end of this book.

2021 - Office of National Intelligence, John Ratcliffe

Former Director of the Office of National Intelligence, John Ratcliffe said in a recent interview on FOX news "Frankly, there are a lot more sightings than have been made public" and "there are instances where we don't have good explanations for some of the things that we've seen," and 'when that information becomes declassified, I'll be able to talk a little bit more about that."

Ratcliffe indicated "We are talking about objects that have been seen by Navy or Air Force pilots, or have been picked up by satellite imagery that frankly engage in actions that are difficult to explain, movements that are hard to replicate, that we don't have the technology for and are traveling at speeds that exceed the sound barrier without a sonic boom." and "I think the information that is being gathered will be put out in a way the American people can see."

Vice-chair of the Intelligence Committee - Marco Rubio

Sen. Marco Rubio (R-Fla.), Vice-chair of the Intelligence Committee, said the prospect that something otherworldly is behind the flying objects does not concern him as much as the idea that a U.S. adversary could be making secret technological advances.

According to Rubio, “The bottom line is if there are things flying over your military bases and you don’t know what they are because they aren’t yours and they are exhibiting potentially technologies that you don’t have at your own disposal, **that to me is a national security risk** and one that we should be looking into,”

Jun 24, 2021 - Christopher Mellon

In an article titled “Don’t Dismiss the Alien Hypothesis” Chris Mellon stated “The United States has the most advanced aerospace capabilities of any country and spends more than twice as much on defense as Russia and China combined. The Administration and Congress concur that these objects are not classified U.S. aircraft. [18]

In the USS Nimitz incident and other cases some of these vehicles have been observed doing things that we cannot replicate and do not understand. Official records from the U.S. and other countries indicate that this is a global phenomenon that has been occurring since at least WW2 and perhaps far longer.

We have no reason to believe that many of these objects are from Russia or China and in fact it seems improbable. Especially when we consider how long this phenomenon has been observed. Consequently, the “not invented here” hypothesis is the only theory currently consistent with the known facts. From any vantage point we are more likely to encounter probes than signals from ET because probes are far safer, more efficient and more effective for purposes of space exploration. The radical capabilities and mysterious appearance of some of the vehicles being observed are consistent with what we'd expect of probes or craft from a space-faring civilization."

Former CIA Director R. James Woolsey

Former CIA Director R. James Woolsey said he believes UFOs exist and hopes humans would be friendly to Extraterrestrials if they ever make contact according to a report. "What is going on? I don't know. Does anybody know?" he said. "There had just been enough things that have occurred that I think there will be a lot of examination of what's going on over the course of several months or years". He added: "I'm not as skeptical as I was a few years ago, to put it mildly, but something is going on that is surprising to a series of intelligent aircraft, experienced pilots".

Members of the House Intelligence Committee Received a Classified Briefing inside a (SCIF) Sensitive Compartmented Information Facility

June 18, 2021, Capitol Hill lawmakers said Wednesday that UFOs could pose a pressing threat to America's National Security as they emerged from a highly classified briefing with Navy and FBI officials on the unexplained phenomena.

- **Rep. Sean Patrick Maloney (D-NY).**
  "We take the issue of unexplained aerial phenomena seriously to the extent that we're dealing with the safety and security of US military personnel or the National Security interests of the United States so we want to know what we're dealing with," "I think it's important to understand that there are legitimate

questions involving the safety and security of our personnel and in our operations and in our sensitive activities and we all know that there's [a] proliferation of technologies out there," he continued. "We need to understand the space a little bit better."

- **Rep. Val Demings (D-Fla.)**
  Added "You know it's always about our safety and security, our National Security is [priority] number one and so that's really the area where we really focused on this morning."

- **Rep. Adam Schiff (D-Calif.)**
  The committee chairman implied that the briefing was eye opening, but also declined to get into specifics ahead of the release of a public report. Americans should expect an eventual public hearing on the report's findings. "We're looking forward to having a public hearing at some point," he said. "I mean, there's some National Security concerns that we want to take into consideration."

- **Rep. Mike Quigley (D-Ill.)**
  Said that, "if anything, the landmark report shows that UFOs are finally being taken seriously. "The stigma is gone," he said. "Now that's as big a change in policy as I've witnessed about this issue in my lifetime. So the fact that they are taking this sort of thing seriously for the first time, I think, is important."

United Nations Appointing Ambassador to Alien World?

In an article written by Edward Moyer in November 2011 for CBS News, a Malaysian astrophysicist, Mazlan Othman who was director of the United Nations Office for Outer Space Affairs (UNOOSA) will be named as the UN ambassador to extraterrestrials, if and when they contact humanity. UNOOSA deals with space-related issues, ranging from international cooperation in peaceful uses of outer space, to managing the growing problem of space debris.

In the next chapter we're going to take a look at the nine-page Preliminary Assessment Report. It is carefully written and exceptionally vague as far as any concrete details about UAPs, but direct and in fairly simple terms. I would encourage you to read through it. It is reprinted verbatim in the Appendix Section at the end of this book. Then we will do an analysis and discuss it in some detail.

Regardless of what ultimately emerged in this report, the prospect has catalyzed a wave of mainstream coverage of UFO Research, from the New York Times, Washington Post, Fox News and "60 Minutes."

A highly controversial subject is now receiving a surprising level of public respectability as a National Security concern. More politicians are now more comfortable talking about UFOs

Chapter 8
The Assessment Report of 2021
and
The National Defense Authorization Act (NDAA) of 2023

Author's Note: The Consolidated Appropriations Act (H.R. 133) was a $2.3 trillion dollar spending bill that was signed into law on December 27, 2020. It mandated that a report detailing what the government knows about UFOs be released this year. This legislation (over 5600 pages) contained the Intelligence Authorization Act for Fiscal Year 2021 which had a provision titled **"Advanced Aerial Threats."**

This provision mandated that the Director of National Intelligence work with the Secretary of Defense on a report detailing everything the government knows about UFOs (also known as UAPs or AAPs). They were directed to include "detailed analysis of UAP data and intelligence" gathered by the Office of Naval Intelligence, the Unidentified Aerial Phenomena Task Force and the FBI.

When the spending package was signed into law a 180-day countdown began giving intelligence officials until June 25th, 2022, to deliver the report to Congress.

Two Reports Released

A Top Secret Classified version of the Assessment Report, in excess of 70 pages was shared with members of the

House Intelligence Committee in a classified briefing inside a SCIF (Sensitive Compartmented Information Facility) in the U.S. Capital.

Several days later on June 25th the nine-page Assessment Report was released to the public. The public version of the report is very carefully worded, repetitious and exceptionally vague as far as any concrete details about UAPs. This 9-page public Assessment Report really didn't provide a lot of useful information as noted. However, it is the first step in a long process that may ultimately lead to disclosure.

The writers do not acknowledge anything about the origin of UAPs suggesting it might be weather balloons, atmospheric phenomena, foreign adversary systems or even marsh gas, and they would like you to think that maybe the Russians or Chinese might have something to do with this. The only thing they seem to know for sure is that UAPs are a threat to National Security.

This highly controversial subject is now receiving a surprising level of public respectability as a National Security concern. More politicians are now comfortable talking about UFOs. The nine-page Preliminary Assessment report is reprinted verbatim in the Appendix Section of this book for your review.

National Defense Authorization Act 2023

In December 2022, President Biden signed the Fiscal 2023 National Defense Authorization Act (NDAA) into law allotting $816.7 billion dollars to the Defense Department. The act includes provisions to set up a permanent office to investigate alleged UFO / UAP sightings and keep Congress in the loop about what it finds.

Most important, there is language that states, "regardless of any previous written or oral non-disclosure agreements that could be interpreted as a legal constraint on a witness reporting unidentified aerial phenomena," (more commonly known as UFOs) to Congress would not be violating Federal classified information laws if they come forward. **This basically means that people with information about UFO / UAPs have been given amnesty and immunity from prosecution.**

At the time this book is being published whistleblowers have already started coming forward with information about crash wreckage and alien bodies of extraterrestrial origin which have been recovered at crash sites. They are specifically alleging that this crash debris and these bodies have been recovered and stored in the numerous government and defense contractor facilities. Some of this debris is reportedly intact and being used in back engineering projects.

David Grusch is a former National Reconnaissance Officer and Air Force Veteran who served as a high-level U.S. Intelligence Officer, a veteran of the National Geospatial Intelligence Agency, and a member of the Unidentified Anomalous Phenomena Task Force at the Department of Defense. [20]

House Oversight Committee's Subcommittee on National Security (nationally televised hearings).
On Jul 26, 2023, Grusch testified on national television to the House Oversight Committee's Subcommittee on National Security about a top-secret UFO retrieval program and claims of decades-long cover-up of non-human intelligent life by the U.S. Government. The claim is that the secret government program has non-human bodies and wreckage of extraterrestrial origin and the information about it has been kept from the public and Congress for over 50 years.

IGIC (Inspecter General of the Intelligence Committee)
In compliance with the act some whistleblowers have filed complaints about urgent concerns with the IGIC (Inspecter General of the Intelligence Committee) who has ruled this to be an urgent credible concern for Congress.

After investigating the complaints and information that has been provided, in January 2024 the Inspector General (IGIC) conducted closed classified hearings with a number of members of Congress in a SCIF at the US capital.

Although the members of Congress could not specifically talk about the classified information that was disclosed to them, many of them felt that the information provided by David Grusch was accurate and authenticated.

As a result of this briefing the Inspector General is going to be scheduling more hearings with the House Oversight Committee which will hopefully be televised.

In the coming months we want to keep an eye out for David Gresh and some of the other whistleblowers for important developments that will be forthcoming from them.

END CHAPTER

Chapter 9
Stairway to Full Disclosure

Catastrophic Disclosure

The concept of catastrophic disclosure involves the unplanned and potentially chaotic release of earth-shattering UAP revelations which would result in profound geopolitical, social, and economic turmoil that could potentially wreck the stock markets and leave the world economy in mayhem.

Former deputy assistant secretary of defense for intelligence Christopher Mellon stated "a strategic, methodical disclosure of the existence of non-human intelligence, while potentially destabilizing could ultimately bear profoundly positive results. . . a controlled disclosure to facilitate scientific study of such remarkable technology is long overdue"

In congress there is political urgency to disclose the UAP conundrum because some high-ranking military and intelligence officials are concerned that both China and Russia are also reverse-engineering craft of "non-human origin", and they could step forward on the world stage to disclose the true origin of these UAPs and also reveal what the US Government and defense contractors have been doing with these secret reverse-engineering projects for at least 50 years, which obviously increases the potential for a technological surprise from foreign adversaries.

Laying this all out in a blatant disclosure with no holds barred for the general public would not be an easy transition and hopefully won't happen as discussed in the following hypothetical illustration.

War of the Worlds - Season Two - Coming Soon!

Consider a Sunday evening news documentary like CBS 60 minutes, on a nationally televised broadcast talking about UFOs and extraterrestrials.

As the broadcast opens the commentator (one of the best-known personalities) begins by telling the viewers that UFOs are real based on recent high level verified government information that has been given to the network. That they are of extraterrestrials origin and piloted by intelligent entities from various off planet civilizations. (which is actually true).

As the television program continues viewers learn that these extraterrestrial entities have technology that is hundreds of years superior to ours and they have powerful weapons that we've never even heard of yet. (which is also true).

The commentator goes on by explaining that ETs are capable of interdimensional travel, and they can transcend time by traveling from future points. The commentator appears to be upset and nervously talks about Individuals who have had close encounters, who report these entities can read thoughts and telepathically communicate with them. (which is also true).

The broadcast gets darker as discussion about people, traveling in vehicles on remote country roads at night are being stopped (vehicles disabled) and then abducted, as well as people being abducted from their homes in the middle of the night. What is more unthinkable is the scenario where hikers are snatched from national parks, leaving only hiking boots and other articles of clothing behind.

The program ends by suggesting UFOs are thought to be responsible for power blackouts and near misses with aircraft from both military and commercial aviation. Then finally concluding about risks to national security by informing viewers about UFOs in restricted airspace over military bases and facilities that house nuclear weapons. As the program ends viewers are terrified, and panic ensues across the country.

Obviously, a program like this would be devastating, causing panic and chaos like it did in 1938 with the "War of the Worlds" radio drama by Orson Welles. The element of fear is a great controlling factor not to mention the adverse effect on the public confidence in government.

Concept of a Stairway for Disclosure

I used to think about the corporate sales process as similar to a stairway where you start at the bottom on the first step with the basics by evaluating the situation and looking at potential problems and obstacles. Then you develop the client's confidence and help them understand the concepts; as they learn you move up the steps to the next levels where solutions and plans are outlined for them ultimately reaching the top of the stairway where the goals are accomplished.

The point being that you cannot take someone on the bottom step and drag them up to the top step without going through the whole learning process. You must begin at the bottom step, understanding and evaluating the situation, looking at potential complications then teaching advanced concepts for resolution of issues which ultimately relieves the anxiety, taking one step at a time to develop confidence and insight reaching the top step.

Public Change in Perception of Reality

So, in the steps along the way to full disclosure the public is going to have to go through a very difficult learning curve that deals with some very high strangeness issues that we've discussed with the UFO scenarios.

The education process to understand multidimensional consciousness and accept that altered states of reality do exist may be painful but will not seem so bad when the benefits to humanity about miraculous advances in medical technology, free energy production, and technological developments involving anti-gravity transportation become available to everyone.

Longer-Term Disclosure

Since the original draft of this book the stairway is suddenly becoming much steeper as things are developing quickly in congress to initiate change in public awareness and ultimately disclose the full truth about UFOs and their extraterrestrial origin.

The defense contractors and the intelligence agencies would like to slow things down as far as disclosing information, while keeping their fingerprints off most of it as long as possible. Not to mention that the more graduated release of information gives those executives involved with the Military Industrial Complex time to dump their stock, sell their patent rights, take their immense profits, and get the hell out of town before the public realizes what's been going on for the last 70 years.

END CHAPTER

EPILOGUE

The biggest event in human history will be the discovery that we are not alone and that there are other living intelligent entities in this universe that we may learn something from!

The concept of full disclosure for the public, making available direct intimate contact with extraterrestrial civilizations would obviously revolutionize life on Earth.

Not only by Introducing amazing transportation technology, free energy production and unheard of medical innovations but also producing advanced knowledge, enlightenment and an elevation in human consciousness, which will surely break the bonds that the Intellectual Elite and World Bankers have over our Human Society.

So when they show themselves to us, which may be very soon, perhaps we should embrace them and be nice!

END CHAPTER

References

[1] **Largest UFO reporting centers in the USA**:

- **THE NATIONAL UFO REPORTING CENTER** http://www.nuforc.org/ and
- **MUFON (Mutual UFO Network)** http://www.mufon.com/reportufo.html

[2] **Embry-Riddle Aeronautical University** is a private university in the US specializing in aviation and aerospace engineering. It teaches the science, practice, and business of aviation and aerospace, as well as flight training, ground school, and technical training programs. Located in Daytona Beach Florida For more about the University information see http://www.erau.edu/

[3] **CERN** The world's largest and highest-energy particle collider was built by **the European Organization for Nuclear Research (CERN)** between 1998 and 2008 in collaboration with over 10,000 scientists and hundreds of universities and laboratories. It lies in a tunnel 17 miles in circumference and 574 ft deep beneath the France Switzerland Border near Geneva.

[4] **The Mutual UFO Network** (MUFON) is an American non-profit organization that investigates cases of reported UFO sightings. It is one of the oldest and largest civilian UFO investigative organizations in the world. For more information see http://www.mufon.com/

[5] Albemarle County Sheriff's Office Reserve Division The Reserve Division provides law enforcement back up and support for all Albemarle County Sheriff's Office functions including community support activities, providing many valued services at little to no cost to the taxpayer. For more information see http://www.albemarleso.org/Nreserve.html

[6] Central Shenandoah Criminal Justice Training Academy is a State Law Enforcement Training Facility located in Weyers Cave, Virginia, one of ten regional academies within the State of Virginia preparing individuals for

[7] Bigelow Aerospace https://bigelowaerospace.com/ From Wikipedia, the free encyclopedia, Bigelow Aerospace https://en.wikipedia.org/wiki/Bigelow_Aerospace

[8] CBS "60 MINUTES" TV show aired May 28, 2017. Correspondent Lara Logan's interview with Robert Bigelow **"Bigelow Aerospace founder says commercial world will lead in space"** http://www.cbsnews.com/news/bigelow-aerospace-founder-says-commercial-world-will-lead-in-space/

[9] The Day After Roswell

by Philip J. Corso and William J. Birnes

The Day After Roswell is an American book about extraterrestrial spacecraft and the Roswell UFO incident. It was written by United States Army Colonel Philip J. Corso, with help from William J. Birnes, and was published as a tell-all memoir by Pocket Books in 1997, a year before Corso's death. The book claims that an extraterrestrial spacecraft crashed near Roswell, New Mexico, in 1947 and was recovered by the United States government who then sought to cover up all evidence of extraterrestrials. The book appeared on The New York Times Best Sellers List for several weeks,

[10] "**Sent by Eisenhower to Meet an ALIEN**, an Amazing Confession of Ex Military" by Richard Dolan https://www.youtube.com/watch?v=d4O89Xtt47Q

[11] Project Blue Book From Wikipedia, the free encyclopedia, https://en.wikipedia.org/wiki/Project_Blue_Book

[12] NORAD From Wikipedia, the free encyclopedia, https://en.wikipedia.org/wiki/North_American_Aerospace_Defense_Command

[13] False Flag From Wikipedia, the free encyclopedia **https://en.wikipedia.org/wiki/False_flag**

[14] Vallée, J. F. and Davis, E. W., "Incommensurability, Orthodoxy and the Physics of High Strangeness: A 6-Layer Model for Anomalous Phenomena," in Proceedings of the 2nd International Forum on Science, Religion and Consciousness Oct. 2003, eds. J. Fernandes and N. L. Santos, Center for Interdisciplinary Study of Consciousness, University Fernando Pessoa, Porto, Portugal, 2005, pp. 225-239. See full text http://www.jacquesvallee.com/ both in English and French. Follow the link to "personal research" https://www.jacquesvallee.net/wp-content/uploads/2018/11/Incommensurability_Orthodoxy_and_the_Phy.pdf

[15] COMETA - French Report on UFOs and Defense:
A Summary By Gildas Bourdais cometa.pdf (cufos.org)
July 16, 1999 an important document was published in France entitled, UFOs and Defense: What must we be prepared for? ("Les Ovni Et La Defense: A quoi doit-on se préparer?"). This ninety-page report is the result of an in-depth study of UFOs, covering many aspects of the subject, especially questions of national defense. The study was carried out over several years by an independent group of former "auditors" at the Institute of Advanced Studies for National Defense, or IHEDN, and by qualified experts from various fields. Before its public release, it has been sent to French President Jacques Chirac and to Prime Minister Lionel Jospin

[16] Washington Post / New York times December 2017

- **Glowing Auras and 'Black Money': The Pentagon's Mysterious U.F.O. Program**
  https://www.nytimes.com/2017/12/16/us/politics/pentagon-program-ufo-harry-reid.html
  Note: video is embedded in this article that shows an encounter between a Navy F-18 Super Hornet and an unknown object. It was released by the Defense Department's Advanced Aerospace Threat Identification Program.
- **Head of Pentagon's secret 'UFO' office sought to make evidence public**
  https://www.washingtonpost.com/world/national-security/head-of-pentagons-secret-ufo-office-sought-to-make-evidence-public/2017/12/16/90bcb7cc-e2b2-11e7-8679-a9728984779c_story.html

[17] Daily Mail: The Office of Global Access (OGA) - a wing of the CIA - has played a central role in collecting alien spacecraft since 2003, sources tell DailyMail.com https://www.dailymail.co.uk/news/article-12796167/CIA-secret-office-UFO-retrieval-missions-whistleblowers.html

[18] Christopher Mellon - Jun 24 - Article
Don't Dismiss the Alien Hypothesis

https://www.christophermellon.net/post/don-t-dismiss-the-alien-hypothesis

[29] David GRUSCH speaks to the House Oversight Committee's Subcommittee on National Security Jul 26, 2023

https://www.youtube.com/watch?v=reQUctvwcw8

Appendix

STATEMENT TO CONGRESS
by George Knapp

Rep. Burchett and members of the committee,

Thank you for inviting me to share some information that hopefully will be useful to your pursuit of UAP/UFO transparency. The public appreciates your courage in tackling this still controversial subject, especially in light of the inevitable pushback members of Congress have already faced for daring to ask basic questions and for refusing to accept the stonewalling, veiled threats, and overt ridicule that have characterized the position of our military and intelligence agencies for the past 75 years.

My name is George Knapp. I am the chief investigative reporter for KLAS 1V in Las Vegas. KLAS is Nevada's original television station and a CBS affiliate.

As a journalist, my interest in UFO secrecy began in 1987. In the years since then, I have written hundreds of UFO-related news stories and series, probably more stories over a longer period of time than any other mainstream journalist in the country. (For the past 22 years, I have worked with photographer Matt Adams on these news reports.)

From the beginning, it was apparent the subject has a steep learning curve. Initially, I was far more interested in the government's response to UFO questions than I was in the larger questions about intelligence life other than humans. The statements issued by various military spokespersons and government agencies were directly at odds with what those same military officials and intelligence agencies confided to each other behind the scenes. Since 1969, the position of our military has been that UFOs pose no threat to national security and are not worthy of further study. This dismissive attitude is directly at odds with what was revealed in documents, reports, and internal memos. High ranking military officers considered the UFO mystery to be "serious business." The paper trail revealed via FOIA requests documents how military leaders knew that UFOs were "real, not fictitious", that they were metallic craft, capable of incredible maneuvers far beyond any known technology on Earth, and that there were a disturbing number of incidents wherein UFOs seemed to demonstrate an intense interest in our military assets, in particular nuclear weapons. If this is not a matter of national security, what is? Yet the public was assured, again and again, there is really nothing to the UFO subject.

In 1989, I started hearing seemingly-outlandish tidbits regarding crashed saucers, strange materials, and reverse engineering programs being carried out in secrecy in the Nevada desert by intelligence operatives and defense contractors. The first person I told about this-outside of our newsroom - was U.S. Senator Harry Reid, then in his first term in the Senate. Reid said he was interested in hearing more, and that began a private, two-way conversation that continued for the next three decades. I kept Sen. Reid in the loop about UFO developments I was pursuing as a journalist, and he helped me obtain information that I might otherwise be

nable to access on my own. That private conversation with Reid proved to be pivotal in the long run, and is directly related to the current explosion of public interest in the UFO/UAP controversy, including the inquiries now underway in both houses of Congress.

The other person I met in 1989 related to the UFO mystery was a billionaire businessman named Robert Bigelow, who owned vast real estate and hotel properties in multiple states. Bigelow began funding private UFO investigations as well the work of UFO organizations UFO. To date, he has spent more of his own money on UFO investigations than any person in the history of the world-tens of millions of dollars. In 1996, Bigelow created his own research organization, the National Institute for Discovery Science, with a science advisory board made up of PhD-level academics, two of the former astronauts who had walked on the moon, and several scientists who had worked for or consulted with the U.S. military and intelligence agencies. While working on the inside, they had learned bits and pieces about classified UFO investigations and special access programs. The NIDS advisory board members put their careers and reputations at risk in order to pursue truth and transparency.

After NIDS began its own investigations and projects, I informed Senator Reid about the organization and arranged an introduction. Reid attended one of the first board meetings of NIDS and was impressed by the professionalism he witnessed and the pedigrees of the world class Science Advisory Board. The connection that was made between Reid and Bigelow proved to be fortuitous in many ways and is directly related to the inquiry you have launched 27 years later.

One of the topics that was of interest to Reid, Bigelow, and NIDS was Russia's ongoing interest in UFOs. In 1993, there was a brief window of opportunity in the former USSR. The period known as Glasnost offered the possibility that western journalists and researchers might be able to learn about subjects that were previously off limits during the worst days of the Cold War. With the assistance off former US Congressman Jim Bilbray, I met a Russian physicist and national security advisor who was in the U.S. in order to speak about arms control issues and nuclear detente at our national laboratories and nuclear weapons facilities. I asked Dr. Nikolai Kapranov if he might be willing to find high ranking persons in the former USSR who may have been in a position to know about any secret UFO programs or investigations in Russia and whether any of those persons would agree to meet with me. He did just that. It took 8 months to arrange an itinerary and to obtain a formal invitation to visit Moscow. In the spring of that year, I traveled with two colleagues to Moscow, met and interviewed more than a dozen military officials, intelligence operatives, and scientists who had knowledge of UFO incidents and studies in the former USSR during the Cold War. As we learned, Russian leaders commissioned an unprecedented UFO investigation. The order went out to all units in the vast Russian military empire that any UFO incident or report must be fully investigated, witnesses interviewed, evidence collected and then all of those materials were forwarded to an office inside the Ministry of Defense.

The study lasted a full ten years and was likely the largest UFO investigation ever undertaken. Thousands of case files were accumulated. Nearly all of the witnesses who were interviewed were military personnel. Many of the incidents described to me by the program's director Col. Boris Sokolov were alarming. Sokolov said there had been 45 incidents in which Russian warplanes engaged with UFOs, chased them, even shot at them. In most incidents, the UFOs shot away at unbelievable speeds, but in three incidents, the Russian warplanes were dibbled and crashed. Two of the pilots were killed. After those incidents, the MOD issued a nationwide order that UFOs should be left alone because, in the words of a top Air Defense official, "they may have incredible capacities for retaliation." Col. Sokolov also shared information about an alarming incident at a Russian ICBM base in Ukraine. UFOs appeared over the base, performed astonishing maneuvers in front of stunned eyewitnesses and then somehow took control of the launch system. The missiles aimed at the US were suddenly fired up. Launch control codes were somehow entered, and the base was unable to stop what could have initiated World War 3. Then, just as suddenly, the UFOs disappeared, and the launch-control system shut down.

Upon my return from Moscow, I shared much of this information with NIDS, with Senator Reid, and with a senior staff members for the Senate Intelligence Committee. The Russian MOD had confirmed to me that they were studying UFO cases in the hope that they might understand and eventually duplicate the technology that had allowed the UFO pilots to so thoroughly dominate Russian airspace and weapons systems. The information made lasting impression on Senator Reid and others and became a key factor in a secretive program that was launched a few years later.

As the committee knows, the current wave of public and congressional interest in the UFO/ UAP mystery was kickstarted in December 2017 when a front page story in the New York Times revealed the existence of an unknown, unacknowledged UFO study dubbed AATIP. The Times story claimed that Sen. Reid and two colleagues in the Senate (Daniel Inouye and Ted Stevens) had secured $22 million in black budget funds to investigate UFOs. AATIP was launched from OUSDI, the office of the undersecretary for defense intelligence and was managed by a career counterintelligence officer named Lue Elizondo. Elizondo's statements to the Times, including supporting testimony from Navy aviator David Fravor about the so-called Tic Tac incident in 2004, were the core of the newspapers' blockbuster story. That news piece led to a new wave of media interest in UFOs, prompted private inquiries to Sen. Reid from his former colleagues in Congress, and was largely responsible for the creation by Congress of the UAP Task Force, which later morphed into AARO, the current UFO program approved by Congress.

The NY Times report was accurate on many levels. AATIP as real. Lue Elizondo was the man in charge of it. And it was an investigation of UFO encounters involving US military personnel. But some parts of the story were dead wrong. The $22 million secured by Reid did not fund AATIP. Rather, it went into an entirely different effort managed by the Defense Intelligence Agency. The original program was dubbed AAWSAP, the Advanced Weapons System Application program. The man who initiated and managed the program for DIA was a veteran intelligence analyst and rocket expert named Dr. James Lacatski. The contract for AAWSAP was awarded to a subsidiary of Bigelow Aerospace, owned by Robert Bigelow. And the focus of the study was much broader than the military-only encounters investigated by AATIP.

Years after the Times story, the public and members of Congress still have not learned much about AAWSAP. It was likely the largest UFO study ever conducted with the use of government funds. It began in Sept. 2008 and quickly ramped up. At one point, it employed 50 full time investigators, far more than Project Blue Book or the UAP Task Force, or AARO. The team compiled what might be the largest and most sophisticated UFO data warehouse ever created, with more than 200,000 cases catalogued. The data base included reports from civilian organizations and foreign governments, as well as new investigations conducted by boots-on-the ground teams dispatched by AAWSAP's manager in Las Vegas, Dr. Colm
Kelleher. It was an astonishing effort that also produced more than 100 highly detailed research papers, many of them more than 100 pages long. The very first case investigated by AAWSAP was the Tic Tac incident from 2004. An initial report was compiled by DIA personnel then shared with AAWSAP. A much larger 140-page report, packed with detailed analysis of the Tic Tac and its capabilities, was written by AASWAP scientists and engineers. Neither Congress nor the public has ever seen the Tic Tac report or any of the other 100-plus research papers.

One of the things that led to the demise of AAWSAP was the pursuit of certain exotic materials rumored to exist within special access programs. One condition of the Bigelow contract with DIA was that Bigelow's Aerospace plant in Las Vegas must be engineered so that it could accept, store, and study certain exotic materials. AAWSAP managers believe these materials were collected from sites where unknown aircraft had crashed. When Dr. Lacatski began pressing the issue, seeking access to the exotic materials, he was met with harsh rebukes. The door, in essence, was slammed in his face. And powerful interests began to apply pressure to end AAWSAP. It lasted a mere 27 months before the plug was pulled, instead of a five year operation as planned by DIA.

AAWSAP investigated a wider range of phenomena than mystery craft seen in the sky. Some of the encounters reported by intelligence operatives were downright weird. AAWSAP personnel suspected that the sighting of weird creatures and bizarre phenomena in the proximity of UFO activity might be some sort of unintentional side effect of a technology that is seemingly beyond anything we currently possess. The overall directive guiding the study was that they follow the evidence, no matter where it lead, in order to gain an understanding of the big picture and not to limit the investigation to lights in the sky or fleeting glimpses on radar or other sensors. Had AAWSAP: been allowed to continue, our country might have some answers by now.

It is my honor to bring this information forward to Congress and to clarify some of the misinformation that has been widely circulated. After AAWSAP ended, AATIP was created from its ashes. Lue Elizondo was able to keep the investigation going for a few years until his frustration with a lack of interest on the part of DOD higher ups became too much to bear. His exposure of the existence of AATIP was a major factor in the eventual creation of the UAP Task Force and later, of AARO. The world still has very limited understanding of the important work done for DIA in AAWSAP. The public has seen very little of the excellent work done by the Bigelow team. Hopefully Congress can start asking questions about these and other investigations undertaken into these perplexing issues. We have a right to know.

George Knapp
American Investigative Reporter

Preliminary Assessment: Unidentified Aerial Phenomena

25 June 2021

UNCLASSIFIED

OFFICE OF THE DIRECTOR OF NATIONAL INTELLIGENCE

UNCLASSIFIED REPORT

Page 1

SCOPE AND ASSUMPTIONS

Scope

This preliminary report is provided by the Office of the Director of National Intelligence (ODNI) in response to the provision in Senate Report 116-233, accompanying the Intelligence Authorization Act (IAA) for Fiscal Year 2021, that the DNI, in consultation with the Secretary of Defense (SECDEF), is to submit an intelligence assessment of the threat posed by unidentified aerial phenomena (UAP) and the progress the Department of Defense Unidentified Aerial Phenomena Task Force (UAPTF) has made in understanding this threat.

This report provides an overview for policymakers of the challenges associated with characterizing the potential threat posed by UAP while also providing a means to develop relevant processes, policies, technologies, and training for the U.S. military and other U.S. Government (USG) personnel if and when they encounter UAP, so as to enhance the Intelligence Community's (IC) ability to understand the threat. The Director, UAPTF, is the accountable official for ensuring the timely collection and consolidation of data on UAP. The dataset described in this report is currently limited primarily to U.S. Government reporting of incidents occurring from November 2004 to March 2021. Data continues to be collected and analyzed.

ODNI prepared this report for the Congressional Intelligence and Armed Services Committees. UAPTF and the ODNI National Intelligence Manager for Aviation drafted this report, with input from USD(I&S), DIA, FBI, NRO, NGA, NSA, Air Force, Army, Navy, Navy/ONI, DARPA, FAA, NOAA, NGA, ODNI/NIM-Emerging and Disruptive Technology, DNI/National Counterintelligence and Security Center, and ODNI/National Intelligence Council.

Assumptions

Various forms of sensors that register UAP generally operate correctly and capture enough real data to allow initial assessments, but some UAP may be attributable to sensor anomalies.

UNCLASSIFIED REPORT

Page 2

EXECUTIVE SUMMARY

The limited amount of high-quality reporting on unidentified aerial phenomena (UAP) hampers our ability to draw firm conclusions about the nature or intent of UAP. The

Unidentified Aerial Phenomena Task Force (UAPTF) considered a range of information on UAP described in U.S. military and IC (Intelligence Community) reporting, but because the reporting lacked sufficient specificity, ultimately recognized that a unique, tailored reporting process was

required to provide sufficient data for analysis of UAP events.

- As a result, the UAPTF concentrated its review on reports that occurred between 2004 and 2021, the majority of which are a result of this new tailored process to better capture UAP events through formalized reporting.
- Most of the UAP reported probably do represent physical objects given that a majority of UAP were registered across multiple sensors, to include radar, infrared, electro-optical, weapon seekers, and visual observation.

In a limited number of incidents, UAP reportedly appeared to exhibit unusual flight characteristics. These observations could be the result of sensor errors, spoofing, or observer misperception and require additional rigorous analysis.

There are probably multiple types of UAP requiring different explanations based on the range of appearances and behaviors described in the available reporting. Our analysis of the data supports the construct that if and when individual UAP incidents are resolved they will fall into one of five potential explanatory categories: airborne clutter, natural atmospheric phenomena, USG or U.S. industry developmental programs, foreign adversary systems, and a catchall "other" bin.

UAP clearly pose a safety of flight issue and may pose a challenge to U.S. national security. Safety concerns primarily center on aviators contending with an increasingly cluttered air domain. UAP would also represent a national security challenge if they are foreign adversary collection platforms or provide evidence a potential adversary has developed either a breakthrough or disruptive technology.

Consistent consolidation of reports from across the federal government, standardized reporting, increased collection and analysis, and a streamlined process for screening all such reports against a broad range of relevant USG data will allow for a more sophisticated analysis of UAP that is likely to deepen our understanding. Some of these steps are resource-intensive and would require additional investment.

UNCLASSIFIED REPORT

Page 3

AVAILABLE REPORTING LARGELY INCONCLUSIVE

Limited Data Leaves Most UAP Unexplained...

Limited data and inconsistency in reporting are key challenges to evaluating UAP. No standardized reporting mechanism existed until the Navy established one in March 2019. The Air Force subsequently adopted that mechanism in November 2020, but it remains limited to USG reporting. The UAPTF regularly heard anecdotally during its research about other observations that occurred but which were never captured in formal or informal reporting by those observers.

After carefully considering this information, the UAPTF focused on reports that involved UAP largely witnessed firsthand by military aviators and that were collected from systems we considered to be reliable. These reports describe incidents that occurred between 2004 and 2021,

with the majority coming in the last two years as the new reporting mechanism became better known to the military aviation community. We were able to identify one reported UAP with high confidence. In that case, we identified the object as a large, deflating balloon. The others remain unexplained.

- 144 reports originated from USG sources. Of these, 80 reports involved observation with multiple sensors.
- Most reports described UAP as objects that interrupted pre-planned training or other military activity.

UAP Collection Challenges

Sociocultural stigmas and sensor limitations remain obstacles to collecting data on UAP. Although some technical challenges—such as how to appropriately filter out radar clutter to ensure safety of flight for military and civilian aircraft—are longstanding in the aviation community, while others are unique to the UAP problem set.

• Narratives from aviators in the operational community and analysts from the military and IC describe disparagement associated with observing UAP, reporting it, or attempting to discuss it with colleagues. Although the effects of these stigmas have lessened as senior members of the scientific, policy, military, and intelligence communities engage on the topic seriously in public, reputational risk may keep many observers silent, complicating scientific pursuit of the topic.

• The sensors mounted on U.S. military platforms are typically designed to fulfill specific missions. As a result, those sensors are not generally suited for identifying UAP.

• Sensor vantage points and the numbers of sensors concurrently observing an object play substantial roles in distinguishing UAP from known objects and determining whether a UAP demonstrates breakthrough aerospace capabilities. Optical sensors have the benefit of providing some insight into relative size, shape, and structure.

Radiofrequency sensors provide more accurate velocity and range information.

UNCLASSIFIED REPORT

Page 4

But Some Potential Patterns Do Emerge

Although there was wide variability in the reports and the dataset is currently too limited to allow for detailed trend or pattern analysis, there was some clustering of UAP observations regarding shape, size, and, particularly, propulsion. UAP sightings also tended to cluster around U.S.

training and testing grounds, but we assess that this may result from a collection bias as a result of focused attention, greater numbers of latest-generation sensors operating in those areas, unit expectations, and guidance to report anomalies.

And a Handful of UAP Appear to Demonstrate Advanced Technology

In 18 incidents, described in 21 reports, observers reported unusual UAP movement patterns or flight characteristics.

Some UAP appeared to remain stationary in winds aloft, move against the wind, maneuver abruptly, or move at considerable speed, without discernable means of propulsion. In a small number of cases, military aircraft systems processed radio frequency (RF) energy associated with UAP sightings.

The UAPTF holds a small amount of data that appear to show UAP demonstrating acceleration or a degree of signature management. Additional rigorous analysis are necessary by multiple teams or groups of technical experts to determine the nature and validity of these data. We are conducting further analysis to determine if breakthrough technologies were demonstrated.

UAP PROBABLY LACK A SINGLE EXPLANATION

The UAP documented in this limited dataset demonstrate an array of aerial behaviors, reinforcing the possibility there are multiple types of UAP requiring different explanations. Our
analysis of the data supports the construct that if and when individual UAP incidents are resolved they will fall into one of five potential explanatory categories: airborne clutter, natural atmospheric phenomena, USG or industry developmental programs, foreign adversary systems, and a catchall "other" bin. With the exception of the one instance where we determined with high confidence that the reported UAP was airborne clutter, specifically a deflating balloon, we currently lack sufficient information in our dataset to attribute incidents to specific explanations.

Airborne Clutter: These objects include birds, balloons, recreational unmanned aerial vehicles (UAV), or airborne debris like plastic bags that muddle a scene and affect an operator's ability to identify true targets, such as enemy aircraft.

Natural Atmospheric Phenomena: Natural atmospheric phenomena includes ice crystals, moisture, and thermal fluctuations that may register on some infrared and radar systems.

USG or Industry Developmental Programs: Some UAP observations could be attributable to developments and classified programs by U.S. entities. We were unable to confirm, however, that these systems accounted for any of the UAP reports we collected.

Foreign Adversary Systems: Some UAP may be technologies deployed by China, Russia, another nation, or a non-governmental entity.

UNCLASSIFIED REPORT

Page 5

Other: Although most of the UAP described in our dataset probably remain unidentified due to limited data or challenges to collection processing or analysis, we may require additional scientific knowledge to successfully collect on, analyze and characterize some of them. We would group such objects in this category pending scientific advances that allowed us to better understand them. The UAPTF intends to focus additional analysis on the small number of cases where a UAP appeared to display unusual flight characteristics or signature management.

UAP THREATEN FLIGHT SAFETY AND, POSSIBLY, NATIONAL SECURITY

UAP pose a hazard to safety of flight and could pose a broader danger if some instances represent sophisticated collection against U.S. military activities by a foreign government or demonstrate a breakthrough aerospace technology by a potential adversary.

Ongoing Airspace Concerns

When aviators encounter safety hazards, they are required to report these concerns. Depending on the location, volume, and behavior of hazards during incursions on ranges, pilots may cease their tests and/or training and land their aircraft, which has a deterrent effect on reporting.

- The UAPTF has 11 reports of documented instances in which pilots reported near misses with a UAP.

Potential National Security Challenges

We currently lack data to indicate any UAP are part of a foreign collection program or indicative of a major technological advancement by a potential adversary. We continue to monitor for evidence of such programs given the counter-intelligence challenge they would pose, particularly
as some UAP have been detected near military facilities or by aircraft carrying the USG's most advanced sensor systems.

EXPLAINING UAP WILL REQUIRE ANALYTIC, COLLECTION AND RESOURCE INVESTMENT

Standardize the Reporting, Consolidate the Data, and Deepen the Analysis In line with the provisions of Senate Report 116-233, accompanying the IAA for FY 2021, the UAPTF's long-term goal is to widen the scope of its work to include additional UAP events documented by a broader swath of USG personnel and technical systems in its analysis. As the dataset increases, the UAPTF's ability to employ data analytics to detect trends will also improve. The initial focus will be to employ artificial intelligence/machine learning algorithms to cluster and recognize similarities and patterns in features of the data points. As the database accumulates information from known aerial objects such as weather balloons, high-altitude or super-pressure balloons, and wildlife, machine learning can add efficiency by pre-assessing UAP reports to see if those records match similar events already in the database.

- The UAPTF has begun to develop interagency analytical and processing workflows to ensure both collection and analysis will be well informed and coordinated.

UNCLASSIFIED REPORT

Page 6

The majority of UAP data is from U.S. Navy reporting, but efforts are underway to standardize incident reporting across U.S. military services and other government agencies to ensure all relevant data is captured with respect to particular incidents and any U.S. activities that might be relevant. The UAPTF is currently working to acquire additional reporting, including from the U.S. Air Force (USAF), and has begun receiving data from the Federal Aviation Administration
(FAA).

- Although USAF data collection has been limited historically the USAF began a six-month pilot program in November 2020 to collect in the most likely areas to
encounter UAP and is evaluating how to normalize future collection, reporting, and analysis across the entire Air Force.
- The FAA captures data related to UAP during the normal course of managing air traffic operations. The FAA generally ingests this data when pilots and other airspace users report unusual or unexpected events to the FAA's Air Traffic Organization.
- In addition, the FAA continuously monitors its systems for anomalies, generating additional information that may be of use to the UAPTF. The FAA is able to isolate data of interest to the UAPTF and make it available. The FAA has a robust and effective outreach program that can help the UAPTF reach members of the aviation community to highlight the importance of reporting UAP.

Expand Collection

The UAPTF is looking for novel ways to increase collection of UAP cluster areas when U.S. forces are not present as a way to baseline "standard" UAP activity and mitigate the collection bias in the dataset. One proposal is to use advanced algorithms to search historical data captured
and stored by radars. The UAPTF also plans to update its current interagency UAP collection strategy in order bring to bear relevant collection platforms and methods from the DoD and the IC.

Increase Investment in Research and Development

The UAPTF has indicated that additional funding for research and development could further the future study of the topics laid out in this report. Such investments should be guided by a UAP Collection Strategy, UAP R&D Technical Roadmap, and a UAP Program Plan.

UNCLASSIFIED REPORT

Page 7

APPENDIX A - Definition of Key Terms

This report and UAPTF databases use the following defining terms:

Unidentified Aerial Phenomena (UAP): Airborne objects not immediately identifiable. The acronym UAP represents the broadest category of airborne objects reviewed for analysis.

UAP Event: A holistic description of an occurrence during which a pilot or aircrew witnessed (or detected) a UAP.

UAP Incident: A specific part of the event.

UAP Report: Documentation of a UAP event, to include verified chains of custody and basic information such as the time, date, location, and description of the UAP. UAP reports include Range Fouler1 reports and other reporting.

1 U.S. Navy aviators define a "range fouler" as an activity or object that interrupts pre-planned training or other military activity in a military operating area or restricted airspace.

UNCLASSIFIED REPORT

Page 8

APPENDIX B – Senate Report Accompanying the Intelligence Authorization Act for Fiscal Year 2021

Senate Report 116-233, accompanying the Intelligence Authorization Act for Fiscal Year 2021, provides that the DNI, in consultation with the SECDEF and other relevant heads of USG Agencies, is to submit an intelligence assessment of the threat posed by UAP and the progress

the UAPTF has made to understand this threat.

The Senate Report specifically requested that the report include:

1. A detailed analysis of UAP data and intelligence reporting collected or held by the Office of Naval Intelligence, including data and intelligence reporting held by the UAPTF;

2. A detailed analysis of unidentified phenomena data collected by:

a. Geospatial Intelligence.

b. Signals Intelligence.

c. Human Intelligence.

d. Measurement and Signatures Intelligence.

3. A detailed analysis of data of the Federal Bureau of Investigation, which was derived

from investigations of intrusions of UAP data over restricted U.S. airspace;

4. A detailed description of an interagency process for

ensuring timely data collection and centralized analysis of all UAP reporting for the Federal Government, regardless of which service or agency acquired the information;

5. Identification of official accountable for the process described in paragraph 4.

6. Identification of potential aerospace or other threats posed by the UAP to national security, and an assessment of whether this UAP activity may be attributed to one or more foreign adversaries;

7. Identification of any incidents or patterns that indicate a potential adversary, have achieved breakthrough aerospace capabilities that could put U.S. strategic or conventional forces at risk;

8. Recommendations regarding increased collection of data, enhanced research and development, additional funding, and other resources.

UNCLASSIFIED REPORT

Page 9

End Report

Related Government Documents

DEPUTY SECRETARY OF DEFENSE
1010 DEFENSE PENTAGON
WASHINGTON, DC 20301-1010

JUN 25 2021

MEMORANDUM FOR SENIOR PENTAGON LEADERSHIP
COMMANDERS OF THE COMBATANT COMMANDS
DEFENSE AGENCY AND DOD FIELD ACTIVITY DIRECTORS

SUBJECT: Unidentified Aerial Phenomena Assessments

A recent report from the Office of the Director of National Intelligence (ODNI) highlights the current challenges associated with assessing Unidentified Aerial Phenomena (UAP) occurring on or near DoD training ranges and installations. It is critical that the United States maintain operations security and safety at DoD ranges. To this end, it is equally critical that all U.S. military aircrews or government personnel report whenever aircraft or other devices interfere with military training. This includes the observation and reporting of UAPs.

The report also confirmed that the scope of UAP activity expands significantly beyond the purview of the Secretary of the Navy, who heads the Unidentified Aerial Phenomena Task Force (UAPTF), and suggested process improvements to ensure timely collection of consistent data on UAP. Consistent with these recommendations and to improve partnership with the ODNI and other non-DoD organizations, I direct the Office of the Under Secretary of Defense for Intelligence and Security to develop a plan to formalize the mission currently performed by the UAPTF. The plan should:

1. Establish procedures to synchronize collection, reporting and analysis on the UAP problem set, and to establish recommendations for securing military test and training ranges.

2. Identify requirements for the establishment and operation of the new activity, to include the organizational alignment, resources and staffing required, as well as any necessary authorities and a timeline for implementation.

3. Be developed in coordination with the Principal Staff Assistants, the Chairman of the Joint Chiefs of Staff, the Secretaries of the Military Departments, and the Commanders of the Combatant Commands and with the DNI and other relevant interagency partners.

All members of the Department will utilize these processes to ensure that the UAPTF, or its follow-on activity, has reports of UAP observations within two weeks of an occurrence.

INSPECTOR GENERAL
DEPARTMENT OF DEFENSE
4800 MARK CENTER DRIVE
ALEXANDRIA, VIRGINIA 22350-1500

May 3, 2021

MEMORANDUM FOR DISTRIBUTION

SUBJECT: Evaluation of the DoD's Actions Regarding the Unidentified Aerial Phenomena (Project No. D2021-DEV0SN-0116.000)

We plan to begin the subject evaluation in May 2021. The objective of this evaluation is to determine the extent to which the DoD has taken actions regarding Unidentified Aerial Phenomena (UAP). We may revise the objective as the evaluation proceeds, and we will consider suggestions from management for additional or revised objectives.

We will perform the evaluation at the Offices of the Secretary of Defense, Military Services, Combatant Commands, Combat Support Agencies, Defense Agencies, and the Military Criminal Investigative Organizations. We may identify additional locations during the evaluation.

Please provide us with a point of contact for the evaluation within **5 days** of the date of this memorandum. The point of contact should be a Government employee or Military Service Member—a GS-15, pay band equivalent, or the military equivalent. Send the contact's name, title, grade/pay band, phone number, and e-mail address to [redacted]

You can obtain information about the Department of Defense Office of Inspector General from DoD Directive 5106.01, "Inspector General of the Department of Defense (IG DoD)," April 20, 2012, as amended; and DoD Instruction 7050.03, "Office of the Inspector General of the Department of Defense Access to Records and Information," March 22, 2013. Our website is www.dodig.mil.

If you have any questions, please contact [redacted]

Randolph R. Stone
Assistant Inspector General for Evaluations
Space, Intelligence, Engineering, and Oversight

2

DISTRIBUTION:

SECRETARIES OF THE MILITARY DEPARTMENTS
CHAIRMAN OF THE JOINT CHIEFS OF STAFF
UNDER SECRETARY OF DEFENSE FOR RESEARCH AND ENGINEERING
UNDER SECRETARY OF DEFENSE FOR INTELLIGENCE AND SECURITY
COMMANDER, U.S. CENTRAL COMMAND
COMMANDER, U.S. NORTHERN COMMAND
COMMANDER, U.S. SPECIAL OPERATIONS COMMAND
DIRECTOR, DEFENSE ADVANCED RESEARCH PROJECTS AGENCY
DIRECTOR, DEFENSE INTELLIGENCE AGENCY
DIRECTOR, DEFENSE THREAT REDUCTION AGENCY
DIRECTOR, MISSILE DEFENSE AGENCY
DIRECTOR, NATIONAL GEOSPATIAL-INTELLIGENCE AGENCY
DIRECTOR, NATIONAL RECONNAISSANCE OFFICE
DIRECTOR, NATIONAL SECURITY AGENCY/CENTRAL SECURITY SERVICE
DIRECTOR, DEFENSE TECHNOLOGY SECURITY ADMINISTRATION
AUDITOR GENERAL, DEPARTMENT OF THE NAVY
AUDITOR GENERAL, DEPARTMENT OF THE ARMY
AUDITOR GENERAL, DEPARTMENT OF THE AIR FORCE

IMMEDIATE RELEASE

Statement by Pentagon Press Secretary John Kirby on Unidentified Aerial Phenomena Assessment

JUNE 25, 2021

Statement by Pentagon Press Secretary John Kirby on Unidentified Aerial Phenomena Assessment:

Today the Director of National Intelligence delivered to Congress a preliminary assessment on unidentified aerial phenomena (UAP) and the progress that the Intelligence Community and the Department of Defense UAP Task Force has made in understanding this threat.

Analyzing UAP is a collaborative effort involving many departments and agencies, and the Department thanks the Office of the Director of National Intelligence for leading a collaborative effort to produce this assessment, as well as the other contributing departments and agencies.

Incursions into our training ranges and designated airspace pose safety of flight and operations security concerns, and may pose national security challenges. DOD takes reports of incursions – by any aerial object, identified or unidentified – very seriously, and investigates each one.

The report submitted today highlights the challenges associated with assessing UAP occurring on or near DOD training ranges and installations. The report also identified the need to make improvements in processes, policies, technologies, and training to improve our ability to understand UAP.

To that end, Deputy Secretary of Defense Kathleen Hicks today directed the Office of the Under Secretary of Defense for Intelligence and Security to develop a plan to formalize the mission currently performed by the UAPTF.

Made in the USA
Middletown, DE
01 November 2025